Geoethics

Silvia Peppoloni · Giuseppe Di Capua

Geoethics

Manifesto for an Ethics of Responsibility
Towards the Earth

Silvia Peppoloni 🆔
Istituto Nazionale di Geofisica e
Vulcanologia
Rome, Italy

Giuseppe Di Capua 🆔
Istituto Nazionale di Geofisica e
Vulcanologia
Rome, Italy

ISBN 978-3-030-98046-7   ISBN 978-3-030-98044-3 (eBook)
https://doi.org/10.1007/978-3-030-98044-3

*There are no privileges of peoples
but common duties of coexistence ...
and no one is saved alone.*

Giovanni Semerano (1911–2005)

Foreword

For a long time, humans and their ancestors have had to adapt to the whims of the environment. For millions of years, they suffered heat and cold, chased prey and resources as they moved, followed tracks, rivers and coastlines in search of a place to stop, before leaving again. Then something changed. After several tens of millennia in their continent of origin, Africa, tribes of hunters and gatherers of the species *Homo sapiens* began to colonise the world more rapidly and to exploit its riches with a systematic approach never seen before in a large size primate. Around 60,000 years ago, an African mammal became planetary. Shortly after, it was on the shores of the Arctic Sea, 45,000 years ago, at twenty degrees below zero, hunting and slaughtering woolly mammoths. Wherever it passed, ecosystems disintegrated and biodiversity decreased. The large Australian and American herbivores, which had never known such a skilful, armed and organised predator, disappeared. The domestication of plants and animals would follow the end of the last ice age. Thus, the evolutionary arrow was reversed: humans were forced to adapt to the environment for millions of years, and now it is the environment that must laboriously adapt to us, to our intensive livestock farming, industries, metropolises, dams and mines.

What makes us human? This is the great question of philosophy but also of evolution. The answer lies not only in our invasiveness. Those same dark-skinned African hunter-gatherers who hunted mammoths at the North Pole, a few millennia later, in a completely different ecological context, namely in the humid heat of the tropical Indonesian island of Sulawesi, produced refined and delightful rock art. Long before Chauvet and Lascaux caves in Europe, these humans, like us, devoted time and resources to a symbolic and aesthetic activity that served no purpose for survival. They could afford it. That is what makes us human. We are not only invasive; we are also creative. We imagine worlds in our head. We are an ambivalent species, creator and destroyer, from the very beginning.

Geoethics, so well outlined by Silvia Peppoloni and Giuseppe Di Capua in the following pages, has its roots in this peculiar natural history of humanity. As other species do, but with a unique pervasiveness, we change the world around us to make it more comfortable for our survival. The strategy has worked and has given us increasing prosperity (though never for everyone), social and economic development, scientific and technological progress, as well as a great Darwinian demographic success. We are almost eight billion and have occupied every corner of the planet. But the strategy is ambivalent, like almost everything that is human, because we then have to adapt to a world that we ourselves have modified, impoverished and often distorted. For example, it is now written in all international documents that we must 'adapt' to anthropogenic global warming.

Technically, *Homo sapiens* is a 'niche builder', in the sense that it actively contributes to the alteration of its own ecological, social, cultural and technological niche. Like all niche builders, human beings pass on to the next generation not only genes and ideas, i.e., the mixture of biological and cultural evolution, but also the ecological changes introduced. Our children are natives of global warming and the environmental crisis: it is the evolutionary legacy we have transferred to them. As this is a costly legacy, an environmental debt, young people rightly protest. Therefore, our ecological hereditariness as niche builders generates a problem of generational justice, as well as distributive justice, since those who suffer most from the environmental crisis are those peoples who have contributed least to its recent deterioration.

If nature hands us an ambivalent legacy, then our choices greatly depend on ethical maturity. The monumental niche construction in which we plunge is indeed a dangerous game, requiring responsibility and wisdom, requiring precisely *ethics of the earth,* ethics of the Anthropocene, which is also *ethics for Earth* that coincide with human ethics as part of the history of Earth. For some years now, in the most accredited scientific journals, we have been finding data and appeals on the environment and on the social and health costs of its degradation, that are so heartfelt that they resemble those of the most militant and radical ecological movements. These appeals have largely gone unheeded, because they are uncomfortable and perhaps difficult to grasp, even on a cognitive and emotional level, by *Homo* self-called *sapiens*, whose collective mind prefers the here and now, crushed on the present, unable to make ethical commitments that are not close and tangible but concern subjects far away in space and time. Scientists and humanists agree on this: we are blind to what is happening, numbers and data are not enough to make us really identify with the process, and we have a problem of imagination.

That is why it is so important to build, as it is done in this book, an ethics of human responsibility towards the earth that is transcultural and rational, scientifically based, intrinsically interdisciplinary and multidimensional, an ethics of justice towards the poor of the earth and towards future generations, against all localism, sovereignism and selfish populism. Scientific and technological solutions are, and will be, essential to get out of the blind alley we have got ourselves into, but they will not be enough. Social, economic and behavioural changes are needed; hence, the need for ethical reflection on the values that will have to mark and direct these deeper changes.

Geoethics rightly emphasises responsibility, both individual and collective, against any alibi. There have always been violent atmospheric phenomena, of course: we inhabit an active, unpredictable, living planet, which is not made in our image and likeness. But these phenomena are becoming increasingly extreme and frequent and, therefore, more difficult to contain and manage. In such a situation (generated by us), the lack of land maintenance and of fight against hydrogeological instability, building speculation and criminal irresponsibility will have only one obvious result: almost every rainfall will become a flood, a 'disaster' on the evening news. But what kind of disaster? Not a natural disaster but an entirely human disaster: a disaster of poor foresight, of failure to understand and embrace risk, to prevent uncertain outcomes, to adapt to ongoing climate change. Ultimately, it is an ethical disaster, if our children are in those houses swept away by the current, and then we mourn them by invoking divine punishment or nature as evil stepmother.

The environmental crisis is showing its worst side, and it is presenting us with the bill. Even the Covid-19 pandemic is a terrible cost that we are paying for ecological degradation, deforestation and the shortsighted exploitation of animals. We know this, but we have stopped saying it. So even science, not just nature, can become an alibi. Geoscientists are publicly and sometimes peremptorily asked for certainties, forecasts, and quantifications. The political classes, by now incapable of any farsightedness, are looking for footholds, pursuing deresponsibilizing narratives or even comforting fake news. Geoscientists rightly respond with probabilities, uncertainties, risks, projections, scenarios, protection of geodiversity and prevention. They remind us that Earth sciences are, by their very nature, sciences of interdependencies, dilemmas, complex and non-linear relationships, ambivalences (one of which is highlighted in the book—the market for rare-earth elements to fuel not only our addiction to smartphones but also the most sustainable technologies). Thus, geoethics—as Peppoloni and Di Capua explain so well here—does indeed concern geoscientists and their professional ethics but also political decision-makers, media actors and public opinion.

Technologies can also become an alibi. Appropriately, the authors are perplexed by certain feared interventions of climate and environmental geoengineering because they denote a salvific and instrumental vision of technologies, a deresponsibilising vision that ignores the problems of redistribution of effects and the management of possible unintentional consequences on the Earth system. There is no more time for such diversions. If we spend billions of public money and waste decades to build an embankment to protect a lagoon, by the time it starts to work it will be too late, with respect to the progressive rise of sea level.

Individual and collective responsibility means being accountable for one's actions and their effects. A further element of value and originality of geoethics is its realism. As builders of our planetary niche, cognitively and emotionally we cannot help but be a little anthropocentric, i.e., we cannot leave our point of view, and we retain the right to defend our survival, like any species, and to guarantee its continuity in future generations. But anthropocentrism becomes responsible when it realises that human interests now coincide with those of nature, of which we are a part. It follows that geoethics and ecology are today two great and compelling humanist undertakings.

Padua, Italy   Telmo Pievani
January 2021

Contents

Chapter 1
Introduction

Human actions leave deep traces on the planet; this is an indisputable fact (Daru et al., 2021; Elhacham et al., 2020; Ellis et al., 2021; Frederikse et al., 2020; Head et al., 2021; Jouffray et al., 2020; Ripple et al., 2020, 2021; Rockström et al., 2009). The anthropogenic impact on social-ecological systems is profound and induces often irreversible changes to these complex sets of adaptive interrelationships, in which natural and anthropogenic elements are closely connected (Berkes & Folke, 1998; Ostrom, 2009) and whose dynamics remain difficult to assess and predict due to the many variables at play (physical, chemical, biological variables, often with non-linear evolution) and the interactions and feedback between their constituent parts (De Vos et al., 2019; Preiser et al., 2018). Science is proposing increasingly advanced and effective tools to find solutions to current global ecological problems, but science and technology alone cannot guarantee socially acceptable and environmentally friendly solutions. It is therefore necessary that scientific advances and technological applications be accompanied by a discussion on their ethical and social implications, aimed at defining the boundaries of anthropogenic action capable of responsibly confronting the quality of current and future human life and the conservation of the other biotic and abiotic elements that make up the earth.

In this perspective, the sustainable development of modern societies requires scientists who are experts in the dynamics of the earth (the geoscientists), who are also aware of the ethical and social aspects of their activities and who are able to suggest new, prudent, and forward-looking ways of interacting with nature (Peppoloni et al., 2019).

Earth sciences, or geosciences, comprise a set of basic and applied disciplines whose object of study is the earth, its relationships with other bodies in the universe, its constitution and structure, its functioning, the processes that take place within it and on its surface, and their evolution in relation to time and space, and the close links between all its characteristics and human activities, including the use and management of its resources. Earth is a set of interacting subsystems (atmosphere, hydrosphere, cryosphere, geosphere, biosphere and anthroposphere) which, in turn, result from the union of physical, chemical, and biological but also cultural and technological elements (Ellis, 2017).

For some years now, there has been a growing awareness in the geoscience community that technical and scientific knowledge must be accompanied by philosophical reflection and a practice (individual and social) that considers the interaction between human beings and the Earth system as a specific object of analysis. This analysis can define the best ways of implementing this relationship, in the light of shared values that overcome the differences of the various social-ecological and cultural contexts.

Understanding how Earth works, how to intervene in its natural systems and processes or how to use its resources are activities that imply great responsibility, not only for those who physically operate in the territory, but also for the entire community that lives in that territory and benefits from it. The ways in which we feel such a responsibility also affects the possibility of identifying the most effective actions to live with the natural hazards by which we are all threatened (earthquakes, floods, volcanic eruptions, climate change, rising sea level). It also affects how we cope with other global emergencies such as pollution, habitat destruction and reduction of biodiversity, soil degradation and consumption, uncontrolled exploitation of non-renewable natural resources and, not least, zoonoses. The SARS-CoV-2 pandemic is only one of the tragic effects of the increasing disruption of already unstable balances between human communities and the environment (Allen et al., 2017; Gibb et al., 2020; Morens & Fauci, 2020). Urbanisation and deforestation increase closeness, thus increasing the possibilities of contact between living species, favouring the spillover of infectious agents from one species to another (Morens & Fauci, 2020; Quammen, 2012). At that point viruses and bacteria also begin to spread among humans, especially in situations of increased social vulnerability due to poor sanitary conditions, neglect, and ignorance, or where social and economic inequalities are more pronounced. Historically, infectious pathogens (bacteria and viruses in particular) have always represented one of the greatest threats to human life, and global anthropogenic changes increase this threat. It is undoubtedly true that epidemics have killed millions of people even in times when the anthropogenic impact on the planet and the size of the world's population were much smaller than observed in the last 150 years, but it is also true that, in past centuries, medicine could not avail itself of modern tools for preventing, combating, and treating infections, including vaccines and antibiotics (Gallavotti, 2019).

However, while the global emergency caused by the SARS-CoV-2 pandemic may have weakened the common perception of the importance of environmental issues and the need for a transition to a more sustainable and eco-friendly world economy, it has also increased awareness of the importance of individual behaviour in addressing global threats and personal responsibility in implementing any strategy to reduce the impact of such threats, as well as the need to strengthen international institutions and agreements to cope with global risks. Due to the shutdown of industries and businesses in many nations and the stoppage of international trade during the most critical phases of the pandemic, the global economy entered a crisis phase with consequences more severe than those suffered during the great economic crisis of

2008–2009,[1] with devastating effects on supply chains and on the economies of developing countries, as well as leading to a sharp increase in inequalities.[2] The negative effects of the pandemic have clearly shown the extent to which human communities are interconnected at a planetary level and how this requires us to respond to global threats with common languages, strategies, and actions through international bodies capable of a 360-degree vision of the challenges to be faced. They must be able to plan interventions on the basis of data and knowledge from different parts of the world, to help each country in defining specific local decision-making processes and the best problem-solving strategies based on its own social and cultural peculiarities. Without the creation of international decision-making bodies to deal with planetary emergencies, humanity runs the risk of finding itself in new major systemic crises in the future, an eventuality that is more probable if current generations will be unable to make drastic and effective decisions in reducing global anthropogenic impacts.

In recent years, these concerns and uncertainties have fuelled the need for geoscientists to reflect on the ethical values that underpin their work, and the desire to become more aware of the responsibilities that arise from a commitment to Earth science research and practice. Their reflections have gradually given rise to geoethics (for a more detailed definition see Chap. 2), as a speculative field aimed at identifying and implementing the most appropriate and responsible behaviours and actions in the management of natural processes, redefining the interaction between human beings and the Earth system on the basis of a scientific, critical, pragmatic approach, as distant as possible from rigid ideological positions (Bobrowsky et al., 2017; Bohle, 2019; Peppoloni & Di Capua, 2012, 2015, 2020, 2021; Wyss & Peppoloni, 2015).

Geoethics was born and developed in order to identify the values and ethical criteria that can guide the relationship that binds humans to the earth, through actions able to guarantee a balance between the preservation of the planet's habitability and the economic and social development of our societies (Peppoloni & Di Capua, 2020; Peppoloni et al., 2019), identifying a *safe operating space for humanity* (Rockström et al., 2009).

The ideas that inspire the theoretical foundations of geoethics can be traced back to the nineteenth century, when the anthropogenic impact on nature began to be recognised and documented, and both technical and cultural considerations were formulated. Since then, technological, industrial, and social development, the resulting rapid population growth and urban sprawl have significantly increased anthropogenic impacts on the planet, with damages now evident at all scales.

Geoethics is the realization that the human agents are modifiers, only partly aware of the natural environments and territories in which they operate and live, of their physical, chemical, and biological characteristics, as well as of the social and cultural traits that characterise those territories. Currently, anthropogenic action is carried

---

[1] https://foreignpolicy.com/2020/03/18/coronavirus-economic-crash-2008-financial-crisis-worse/; https://www.csis.org/analysis/global-economic-impacts-covid-19. Accessed 29 March 2022.

[2] https://www.undp.org/content/undp/en/home/news-centre/news/2020/COVID19_Crisis_in_dev eloping_countries_threatens_devastate_economies.html. Accessed 29 March 2022.

out through the use of impressive technological development, which poses further problems and also offers possible solutions. Anthropogenic changes on the planet influence people's economic development and social prospects, and they require us to consider, from different perspectives, crucial issues such as the prudent use of natural resources and energy, defence against natural and anthropogenic risks, pollution reduction, soil conservation, mitigation of global environmental change and the adaptation that such change requires. Hence, there is a need to broaden the scientific horizon by means of philosophical and sociological reflection, exploring issues of equity, intra- and intergenerational justice, access to quality education, and calling for action on a political level based on listening, prudence, wisdom and foresight.

Local and global anthropogenic changes, natural risks, transition towards new economic paradigms, and sustainable development are topics that have gone beyond the scientific sphere and have entered the global public debate. In his encyclical *Laudato si'*, Pope Francis also emphasises the need for an 'ecological conversion' of humanity, which will lead everyone to take responsibility for 'caring for the common home' (Pope Francesco, 2015). In this instance, *'caring'* means that set of actions that cannot be postponed on an environmental, social, and legal level, aimed at increasing attention to the issues of inequality, poverty, the shared availability of natural resources, and the need for international governance of global commons to change current unsustainable development models.

The recent demonstrations by young people in Europe and other parts of the world, calling for more incisiveness on the part of governments in implementing policies to counter global warming, can also be seen as part of this broader phenomenon of generalised concern over the lack of an effective international governance, determined to tackle the major environmental issues with due urgency and a sense of responsibility. The, at best, ambivalent results of the UN Climate Change Conference (COP26), held in Glasgow from 31 October to 12 November 2021, have further frustrated the expectations of these environmental activists and left a sense of discouragement due to the wait-and-see approach by some governments that continue to postpone these admittedly difficult decisions on social, economic, and political levels, but which are now necessary in order to address climate change.

Global changes no longer arouse only unrealistic or anticipated fears but raise concrete alarms that have been followed up by important global organisations for some time, including the United Nations. Among the various initiatives, in 1988 the Intergovernmental Panel on Climate Change (IPCC[3]) was established. It is the main international scientific body whose aim is to provide governments with scientific knowledge on climate change—past, present and future—analysing their impacts on human and ecological systems, studying the vulnerability and adaptability of these systems, assessing climate change mitigation actions and methods of reducing greenhouse gas emissions, and removing these gases already present in the atmosphere. On the political level, the 17 Sustainable Development Goals of the United Nations[4]

---

[3] https://www.ipcc.ch/. Accessed 29 March 2022.

[4] https://sustainabledevelopment.un.org/sdgs. Accessed 29 March 2022.

have now become the programmatic reference framework for setting the government actions of the states, aimed at guaranteeing humanity a future of health, safety, equity and sustainability.

It is evident that any concrete action aimed at healing ecological 'wounds' requires sharing a solid scientific basis and the support of a framework of reference values that can be convincingly put forward even in very different social, economic, and cultural contexts. In fact, problems with global impact cannot be resolved by means of circumscribed local actions which, although legitimised by the historical and cultural diversity of the various populations that make up the complex mosaic of humanity, cannot constitute common and decisive operational models on their own.

Geoethics attempts to overcome this fragmentation by proposing geosciences as the foundation of responsible human action towards the earth. Geoethics is conceived as a rational, operational, and multidisciplinary language, capable of binding and concretely supporting a cohesive international community, committed to the shared resolution of global environmental problems and complex challenges which, in fact, do not know national, cultural or religious boundaries.

However, before addressing society, geoethics requires consciousness-raising, albeit not generalised, by geoscientists themselves. This initial phase of reflection within the geoscience community first concerned professional ethics and then the social and political nature of the geosciences. In this search for the deepest meaning of geological knowledge and the social and cultural value of its scientific results as instruments in the service of the common good, the identification and valorisation of some eminent personalities of the past, ideally considered as forerunners of contemporary geoethical thought, was fundamental.

The current development of the geoethical movement and thought now has an international dimension. As such, it requires the conceptual categories, content, and tools of geoethics be defined in an increasingly articulate way and made available to both the scientific community and the general public, to involve both in the debate on the responsibilities of the human agent. In a more strictly scientific sphere, the criteria and values of geoethics can offer insights and can guide the discussion on the usefulness of geosciences in solving global environmental issues, on the aspects of intra-professional cooperation, and on the fundamental role of geoscientists in the societal contexts in which they operate.

In recent years, geoethical thought has unequivocally highlighted how the daily functioning of modern societies depends on responsible application of geological-environmental knowledge, which can enable us to keep our home, the earth, in the best possible condition for habitability, so as to ensure the life and development of the human species in balance with other living forms. Understanding the dynamics of the planet in order to predict the consequences for life and the environment means rethinking the physical–chemical-biological system of complex relationships, of which humans are an integral part, in the light of new awareness. In this sense, the drama of the SARS-CoV-2 pandemic was at least useful in confirming some key points which also connote geoethical thought (Di Capua & Peppoloni, 2020):

a) In dealing with global crises, individual behaviour makes a difference. At the root of the chain of actions that a community must deploy to resolve its problems, there is always the individual, who is called upon to confront their sense of responsibility towards the human beings closest to them, as well as towards society, understood as the broader sphere of their human relations.

b) Personal, interpersonal, and social responsibilities are fundamental to living healthily and safely in a globalised and highly interconnected society.

c) Everyone's responsibility towards the Earth system implies respect for social-ecological systems. Poorly considered anthropogenic actions have the effect of increasing the exposure of human communities to unpredictable and rapid phenomena, which may jeopardise the current structure of globalised society, leading to a possible systemic collapse, with no possibility of adaptation to the changed conditions.

d) Merit and competence are values that must be placed at the heart of a new social contract between science and citizens. Scientists who are experts on a given subject are called upon to provide reliable answers, even if they are affected by a certain degree of uncertainty. Facing global warming and its consequences requires competence, professional updating, cooperation, honest discussion with colleagues, and open dialogue among those who hold different positions. Decision-makers are asked to make decisions based on science, carefully evaluating their social and economic aspects, and taking into account the environmental context involved.

e) Global supply chains need to be redefined to be more resilient to shocks, to be less ecologically impactful, and to be in line with the principle of distributive justice, in which everyone has the right to fair access to the earth's resources. This transformation process will be inevitable, albeit difficult and slow as it will have to take into account the complexity of social and economic structures and necessarily involve multidisciplinary approaches.

f) The change in economic, social, and political paradigms, which is indispensable for providing a concrete and effective response to global anthropogenic problems, must also be accompanied by a cultural change in society, which will make it possible to understand the need for the process and facilitate its implementation. To this end, it will be essential to invest in the education, university, and research systems.

g) It is essential to create more transparent, authoritative, and independent international governance mechanisms in the field of health and the environment, which encourage the exchange of knowledge and experience between nations and provide decision-making support to governments. These bodies should favour the integration of each country's decisions referring to local contexts into the globalised human system.

The aim of this essay is to stimulate discussion on these issues, written in the conviction that preserving Earth's habitability is increasingly becoming a matter of individual and social responsibility. With this in mind, our 'Manifesto for Geoethics?' aims to help identify the social and cultural conditions that are indispensable for initiating the development of international governance, which ensures human progress based on health, security, justice and harmony.

References

Allen, T., Murray, K. A., Zambrana-Torrelio, C., Morse, S. S., Rondinini, C., Di Marco, M., Breit, N., Olival, K. J., & Daszak, P. (2017). Global hotspots and correlates of emerging zoonotic diseases. *Nature Communications, 8,* 1124. https://doi.org/10.1038/s41467-017-00923-8

Berkes, F., & Folke, C. (Eds.). (1998). *Linking social and ecological systems* (p. 476). Cambridge University Press, ISBN 978–0521785624.

Bobrowsky, P., Cronin, V. S., Di Capua, G., Kieffer, S. W., & Peppoloni, S. (2017). The emerging field of geoethics. In L. C. Gundersen (Ed.), *Scientific integrity and ethics: With applications to the geosciences* (pp. 175–212), American Geophysical Union, Special Publications 73. https://doi.org/10.1002/9781119067825.ch11

Bohle, M. (Ed.). (2019). *Exploring geoethics—Ethical implications, societal contexts, and professional obligations of the geosciences* (p. XIV+214). Palgrave Pivot, Springer International Publishing, ISBN 978–3030120092. https://doi.org/10.1007/978-3-030-12010-8

Daru, B. H., Davies, T. J., Willis, C. G., Meineke, E. K., Ronk, A., Zobel, M., Pärtel, M., Antonelli, A., & Davis, C. C. (2021). Widespread homogenization of plant communities in the Anthropocene. *Nature Communications, 12,* 6983. https://doi.org/10.1038/s41467-021-27186-8

De Vos, A., Biggs, R., & Preiser, R. (2019). Methods for understanding social-ecological systems: A review of place-based studies. *Ecology and Society, 24*(4), 16. https://doi.org/10.5751/ES-11236-240416

Di Capua, G., & Peppoloni, S. (2020). *Il riscaldamento globale al tempo del coronavirus.* MicroMega, La Mela di Newton, https://pikaia.eu/il-riscaldamento-globale-al-tempo-del-coronavirus/. Accessed 11 November 2021.

Elhacham, E., Ben-Uri, L., Grozovski, J., Bar-On, Y. M., & Milo, R. (2020). Global human-made mass exceeds all living biomass. *Nature, 588,* 442–444. https://doi.org/10.1038/s41586-020-3010-5

Ellis, E. C. (2017). Physical geography in the Anthropocene. *Progress in Physical Geography, 41*(5), 525–532. https://doi.org/10.1177/0309133317736424

Ellis, E. C., Gauthier, N., Goldewijk, K. K., Bliege Bird, R., Boivin, N., Diaz, S., Fuller, D. Q., Gill, J. L., Kaplan, J. O., Kingston, N., Locke, H., McMichael, C. N. H., Ranco, D., Rick, T. C., Shaw, M. R., Stephens, L., Svenning, J. –C., & Watson, J. E. M. (2021). People have shaped most of terrestrial nature for at least 12,000 years. *PNAS, 118*(17). https://doi.org/10.1073/pnas.2023483118

Frederikse, T., Landerer, F., Caron, L., Adhikari, S., Parkes, D., Humphrey, V. W., Dangendorf, S., Hogarth, P., Zanna, L., Cheng, L., & Wu, Y. -H. (2020). The causes of sea-level rise since 1900. *Nature, 584,* 393–397. https://doi.org/10.1038/s41586-020-2591-3

Gallavotti, B. (2019). *Le grandi epidemie: Come difendersi* (p. X+198). Donzelli Editore, Roma, ISBN 978–8868438821.

Gibb, R., Redding, D. W., Chin, K. Q., Donnelly, C. A., Blackburn, T. M., Newbold, T., & Jones, K. E. (2020). Zoonotic host diversity increases in human-dominated ecosystems. *Nature, 584,* 398–402. https://doi.org/10.1038/s41586-020-2562-8

Head, M., Steffen, W., Fagerlind, D., Waters, C., Poirier, C., Syvitski, J., Zalasiewicz, J., Barnosky, A. D., Cearreta, A., Jeandel, C., Leinfelder, R., Mcneil, J. R., Rose, N., Summerhayes, C., Wagreich, M., & Zinke, J. (2021). The Great Acceleration is real and provides a quantitative basis for the proposed Anthropocene Series/Epoch. *Episodes*, Online First. https://doi.org/10.18814/epiiugs/2021/021031

Jouffray, J. B., Blasiak, R., Norström, A. V., Österblom, H., & Nyström, M. (2020). The Blue Acceleration: The trajectory of human expansion into the Ocean. *Perspective, 2*(1), 43–54. https://doi.org/10.1016/j.oneear.2019.12.016

Morens, D. M., & Fauci, A. S. (2020). Emerging pandemic diseases: How we got to COVID-19. *Cell, 182.* https://doi.org/10.1016/j.cell.2020.08.021

Ostrom, E. (2009). A General framework for analyzing sustainability of social-ecological systems. *Science, 325*, 419–422. https://doi.org/10.1126/science.1172133

Peppoloni, S., Bilham, N., & Di Capua, G. (2019). Contemporary geoethics within the geosciences. In M. Bohle (Ed.), *Exploring geoethics—Ethical implications, societal contexts, and professional obligations of the geosciences* (pp. 25–70), Palgrave Pivot. https://doi.org/10.1007/978-3-030-12010-8_2

Peppoloni, S., & Di Capua, G. (2012). Geoethics and geological culture: Awareness, responsibility and challenges. *Annals of Geophysics, 55*(3), 335–341. https://doi.org/10.4401/ag-6099

Peppoloni, S. & Di Capua, G. (Eds.) (2015). Geoethics—The role and responsibility of geoscientists. *Geological Society of London, Special Publications 419*, 187, ISBN 978–1862397262. https://doi.org/10.1144/SP419.0

Peppoloni, S., & Di Capua, G. (2020). Geoethics as global ethics to face grand challenges for humanity. In G. Di Capua, P. T. Bobrowsky, S. W. Kieffer & C. Palinkas (Eds.), *Geoethics: Status and Future Perspectives, Geological Society of London, Special Publications, 508*, 13–29. https://doi.org/10.1144/SP508-2020-146

Peppoloni, S., & Di Capua, G. (2021). Geoethics to start up a pedagogical and political path towards future sustainable societies. *Sustainability, 13*(18), 10024. https://doi.org/10.3390/su131810024

Pope Francesco. (2015). *Encyclical Letter LAUDATO SI' of the Holy Father Francis on care for our common home* (p. 184). Vatican Press. https://www.vatican.va/content/dam/francesco/pdf/encyclicals/documents/papa-francesco_20150524_enciclica-laudato-si_en.pdf. Accessed 11 November 2021.

Preiser, R., Biggs, R., De Vos, A., & Folke, C. (2018). Social-ecological systems as complex adaptive systems: Organizing principles for advancing research methods and approaches. *Ecology and Society, 23*(4), 46. https://doi.org/10.5751/ES-10558-23044

Quammen, D. (2012). *Spillover: Animal infections and the next human pandemic* (p. 587). W W Norton & Co Inc, ISBN 978–0393066807.

Ripple, W. J., Wolf, C., Newsome, T. M., Barnard, P., & Moomaw, W. R. (2020). World scientists' warning of a climate emergency. *BioScience, 70*(1), 8–12. https://doi.org/10.1093/biosci/biz088

Ripple, W. J., Wolf, C., Newsome, T. M., Gregg, J. W., Lenton, T. M., Palomo, I., Eikelboom, J., Law, B. E., Huq, S., Duffy, P. B., & Rockström, J. (2021). World scientists' warning of a climate emergency 2021. *BioScience, 71*(9), 894–898. https://doi.org/10.1093/biosci/biab079

Rockström, J., Steffen, W., Noone, K., Persson, A., Chapin, F. S., Lambin, E. F., Lenton, T. M., Scheffer, M., Folke, C., Schellnhuber, H. J., Nykvist, B., de Wit, C. A., Hughes, T., van der Leeuw, S., Rodhe, H., Sörlin, S., Snyder, P. K., Costanza, R., Svedin, U., Falkenmark, M., Karlberg, L., Corell, R. W., Fabry, V. J., Hansen, J., Walker, B., Liverman, D., Richardson, R., Crutzen, P., & Foley, J. A. (2009). A safe operating space for humanity. *Nature, 461*(7263), 472–475. https://doi.org/10.1038/461472a

Wyss, M., & Peppoloni, S. (Eds.). (2015). *Geoethics: Ethical challenges and case studies in earth sciences* (p. 450). Elsevier, ISBN 978–0127999357. https://doi.org/10.1016/C2013-0-09988-4

Chapter 2
Origins of Geoethical Thought

2.1 The Search for Roots

Reflection on the relationship between humans and nature permeates the whole of human cultural experience. Understanding the reality in which individuals live and with which they must enter into a functional relationship is the foundation of the activity of knowledge.

Investigating and getting to know nature, its processes, forms and dynamics, is not an act that is an end in itself, but rather it means assigning spatial and temporal coordinates to reality so that human beings can define scenarios and foresee situations in relation to their possible choices, their actions. This implies that their ethical dimension is projected onto their relationship with nature and conforms their vision, defining the constellation of ideal and experiential references.

At the beginning of the third millennium, geosciences and geoethics are proposed as rational tools necessary to consciously and responsibly address modern issues affecting social-ecological systems.

The geosciences are a highly articulated body of knowledge, both theoretical and practical, for studying and monitoring the causes, effects, and evolution of natural and anthropogenic phenomena, developed largely over the last hundred years and which has become an indispensable support for the functioning of modern societies (Bickford, 2013; BRGM, 2009; GSL, 2014; Wysession & Rowan, 2013).

Geoethics, on the other hand, has received considerable impetus only in the last ten years, although many of the concepts and categories used in its theoretical framework can be found in the reflections of illustrious intellectuals of the past. Among them are Antonio Stoppani (1824–1891), Italian geologist and abbot, author in 1876 of *Il Bel Paese*; Élisée Reclus (1830–1905), French anarchist geographer, author among others of *L'Homme et la Terre*, published posthumously between 1905 and 1908; Aldo Leopold (1887–1948), American ecologist who, in 1949, published his work

© The Author(s), under exclusive license
to Springer Nature Switzerland AG 2022
S. Peppoloni and G. Di Capua, *Geoethics*,
https://doi.org/10.1007/978-3-030-98044-3_2

A Sand County Almanac: And Sketches Here and There; and Felice Ippolito (1915–1997), Italian geologist and engineer, author of the essay *La natura e la storia (Nature and History)*, published in 1968.

The thoughts of these scholars, although with different sensitivities and cultural backgrounds, constitutes the ground on which geoethics has set its reflections on the ethical dimension of the scientist who investigates Earth on the relationship between human being and the planet, on the meaning and societal repercussions of geological knowledge and the practice of the geosciences. Although they have never used the recently coined word 'geoethics', in their philosophical speculations one can identify the theoretical foundations on which geoethical thought has been structured and is evolving, and from which new ideas and suggestions may arise, as well as evidence of its broad cultural potential.

2.1.1   Man Is a 'telluric force': Antonio Stoppani

Known also for his effigy on a famous Italian cheese, the author of *Il Bel Paese* is actually considered one of the fathers of modern Italian geology for the richness of his observations on the natural environments that characterise the Italian peninsula, and he is, perhaps, one of the first geologists to have grasped the profound cultural and educational potential of geological science for human societies. Stoppani believed everyone should begin to know themselves through the study of the natural and physical history of their country (Lucchesi, 2017). He emphasises the pedagogical role of Earth sciences as a tool for the cultural and human growth of society, and he places the link with one's own territory as the foundation of an individual's identity.

In his texts, Stoppani does not limit himself to transmitting observations and geological knowledge necessary to set up that process of inductive analysis that he believes to be the basis of geology. Above all, he wants to suggest an image/idea of nature based on the concepts of sacredness, beauty, and harmony that, although rooted in the naturalistic sensibility of the Stoppani geologist, probably do not prescind from the religious vision of nature of the Stoppani abbot, in any case never connoted by creationist conceptions. A sacred, beautiful, and harmonious nature cannot be irresponsibly destroyed by humans, since it constitutes the substratum of their own human identity, and perhaps in Stoppani's perspective it represents a flash of divine light present in every animate and inanimate natural element. Today, the respect for nature to which Stoppani refers is expressed in scientific terms by the concepts of geological heritage, geodiversity, and geo-conservation, identifying respectively the object of attention of our actions on the abiotic natural environment, its intrinsic and peculiar qualities, and the actions that safeguard its existence even in a context of natural instability.

In Stoppani's vision of nature, and as a result of his observations and practical experience, a human being is an active modifier of the environment (a new *telluric force*), to such an extent that this author introduces into human cultural history the revolutionary concept of the 'Anthropozoic Era', a geological time characterised by the appearance on the planet of a 'recent' geological force, the human agent, capable of profoundly intervening on environmental forms and processes in order to *impose himself on nature*. The concept of the 'Anthropozoic Era' anticipates by more than a century that of the Anthropocene proposed by Stoermer in the 1980s and formulated by Crutzen in 2002 (Crutzen, 2002). Stoppani does not pretend to scientifically formalise a new era by applying the criteria of stratigraphy, nor could he do so in the second half of the nineteenth century, but the enormous cultural significance of his intuition is evident if we consider the time in which it was theorised.

Stoppani knew that the human being of the nineteenth century was not widely aware of the 'geological' power they can wield over Earth (and even today such an awareness is not fully acquired in society). Probably humans themselves, immersed in a positivist cultural climate, do not fully understand that imposing oneself on nature implies a contrast with it on the part of the human being, which is not compatible with the respect for the sacred character of nature itself that Stoppani perceived and to which he recalls us.

2.1.2 'Man Is Nature Becoming Conscious of Itself': Elisée Reclus

The work of this geographer can be summed up in the phrase: 'L'Homme est la nature prenant conscience d'elle même' (Man is nature becoming conscious of itself) (Reclus 1905–1908). In his book *L'Homme et la Terre* this phrase is placed under a drawing depicting Earth held in two hands, an image very similar to the one on the cover of the first volume entirely dedicated to geoethics, published in 2015 (Wyss & Peppoloni, 2015), signifying the responsibility now historically assigned to humanity to care for the planet and life in all its forms. Equally eloquent is another graphic reproduction in Reclus' book, where a man seated between two caryatids, symbolising geography and history, observes the beauty of Earth from afar, as if looking down from a terrace on the Moon.

Here, too, there is a strong resemblance to another iconic image of our time, namely the famous colour photo taken on 7 December 1972, called 'blue marble', in which Earth is observed from space and captured in all its beauty, uniqueness, and fragility by the crew of NASA's *Apollo 17* mission. This photo, which has become a symbol for global environmentalism because of its powerful and nostalgic reference to a potentially 'perfect' home (for some, almost a Spinoza's *natura naturata*), in fact constantly threatened by humans, seems to already exist in Reclus' mind. The 'blue marble' is a finite object and the human being who inhabits it, even the freest and most independent, is bound by its physical, chemical, and biological dynamics.

From this point of view, Reclus appears to be an extraordinary and visionary precursor of geoethical thought. The freedom of the human being passes through the recognition of the bond that binds them to Earth and, consequently, through the epistemological necessity of knowing the laws that regulate the functioning of the planet, to *conform human existence to them* (Battaglia, 2012). It is in the dialectic of human being-Earth, unity-diversity, freedom-limits that humanity develops awareness of its responsibilities towards its own future of freedom, conditioned by respect for the laws and natural balances that must be known scientifically. In line with this thought, scientifically knowing nature and living in contact with it (grasping, in Stoppani's way, its harmony and balance) for Reclus are fundamental actions of a new social education (Battaglia, 2012).

Reclus' geography is, in fact, proposed as a 'guiding science' to found, as Battaglia writes, an ecological ethics of responsibility for the planet in which anthropocentrism is not denied but made responsible. Reclus therefore calls for careful management of the e, a concept that is considered one of the foundations of geoethical thought. According to geoethics, in order to heal ecological 'wounds', rather than considering forms of economic and technological regression, it is necessary to responsibly base human development on geoscientific knowledge and on a geo-biological sensibility capable of grasping the intimate connection between the human being and the physical reality in which they dwell.

The individual retains the absolute centrality of a decisive actor in the evolution of a *changing humanity*, capable of developing what today would be defined as an ecological consciousness, which prefigures a humanism with a naturalistic background (Battaglia, 2012) and recalls the feeling of eco-belonging (Franceschelli, 2008).

2.1.3   'Land Ethic': Aldo Leopold

While Stoppani's and Reclus' thoughts are characterized by anthropocentric traits, Aldo Leopold is considered one of the main proponents of an ecocentric vision. Leopold focuses his reflections on the concepts of biodiversity and wildlife management, according to a multidisciplinary scientific approach. It is aimed at the conservation of nature and, in particular, of its wilderness, as an effect of a process of growth of human ethical sensibility that tends to progressively extend from the interpersonal level to the social level and to the land, limiting personal advantage and contrasting the utilitarian vision of nature conceived at the service of human interests.

For Leopold, the conservation of nature represents 'a state of harmony between men and land' (Leopold, 1949) and human actions must be guided by that *land ethic*, according to which the land must be respected for its intrinsic value as an essential component of human communities. This vision prefigures the modern concept of the social-ecological system, consisting of social matrices and environmental elements that variously interact in space and time. Leopold also believes that it is not possible to think of an existence in which there is no anthropic imprint on nature, but the respect that humans must have for animals, soil, and water is the same due to all the members of the community to which they belong, of which nature is an integral part, endowed with full existential dignity.

2.1.4 The Unity of Scientific and Humanistic Cultures: Felice Ippolito

The direct observation of phenomena is the great value that Ippolito attributes to a science that wants to know effectively, that is truly a science of nature (like physics and geology) and does not remain confined to a dimension of pure abstraction (like mathematics). Knowing nature scientifically has the same meaning as making history, since 'when we want to understand a phenomenon we must keep in mind the whole series of previous events, of which the one under consideration is only the last, the object of our attention at that instant' (Ippolito, 1968, p. 23). Ippolito believes that there are no scientific and humanistic cultures, but that there is only one culture, characterised precisely by a dimension of unity.

This vision also belongs to geoethical thought: having as the object and purpose of one's reflection and action a system of great complexity such as Earth implies taking into account the dimension of unity of human cultural experience, capable of grasping together the numerous interactions present in a global system and of thinking about the most functional ways for humanity to inhabit the planet. The aim of geoethics is to provide an articulate response to complex problems, such as those concerning Earth, because they originate in an extremely structured system of relations-interactions, characterised by a substantially non-linear temporal and spatial evolution and of which we only have a partial knowledge, affected by numerous epistemic uncertainties. Moreover, analysing phenomena on a local scale must not distract us from assessing the broader social-ecological contexts in which they tend to develop and evolve. Taking all these aspects into account, those who possess scientific knowledge must also assume the responsibility of acting ethically, transforming that knowledge into ethical action aimed at the common good.

2.2 Ethical and Social Aspects of the Geosciences

Starting from the 1990s, a reflection on the ethical and social implications of geosciences began to develop within the various geological communities. In particular, Eldridge M. Moores (1938–2018) and E-an Zen (1928–2014), during their tenures as presidents of the Geological Society of America, highlighted some ethical issues at the geoscience-society interface. One of those issues concerns the need to 'link scientific knowledge to society's sense of value - to what is right, what is wrong, what is important' (Zen, 1993, p. 2) —and the consideration that (geo)science is a public enterprise, conducted with public funding, and that for this reason scientists should engage in more dialogue with people, promoting and sharing values that lead us to look to a common future (Bobrowsky et al., 2017). In a condition of increasing anthropogenic impact on ecosystems and geological environments due to population growth, one of the main ethical issues that geoscientists are called to face is precisely the identification and strengthening of their social role (Peppoloni et al., 2019). To understand the extent of population growth, it is enough to consider that at the beginning of the 1990s the population of the planet was 5.5 billion, while today it is about 7.5 billion, an increase of 36% in about 25 years.

To the considerations expressed by Zen in 1993, it is worth adding those of the philosopher Robert Frodeman concerning some aspects of the philosophy of geology (Frodeman, 1995). He defines geology as a historical (in line with Felice Ippolito) and interpretative science, concluding that geological reasoning offers the best model of logical-rational structure for dealing with the problems of the twenty-first century. In fact, geological reasoning is based on some unique perspectives in the intellectual and scientific field that use the magnifying lens of geological time and complex systems. These lenses are a peculiar way of interpreting the interactions of human beings with the Earth system and provide the cultural basis for developing the unique contributions that geoethics can make with respect to environmental ethics and other theoretical fields (Peppoloni & Di Capua, 2020).

On the other hand, Moores (1997) raises an important ethical question concerning geology. He explicitly speaks of a crisis in this discipline after a golden age of fifty years, characterised by the affirmation of the theory of plate tectonics, planetary geology and the substantial contribution of the most advanced techniques of image analysis. This crisis is blamed on poor dissemination of geological knowledge within society despite the public demand for scientific information, on the progressive spread of anti-scientific positions, up to real fundamentalist attitudes towards key geological concepts such as the 'deep time' of the Scottish geologist James Hutton (1726–1797), or major issues such as the use of natural resources and fossil fuels. This crisis is attributed also to excessive fragmentation of the international geoscience community, which is incapable of expressing in a common language coherent and shared positions to be clearly proposed to society.

Moores urges the scientific community to become aware of the significant and tangible benefits that geological activities bring to society and to understand why the geosciences are perceived as socially irrelevant disciplines, despite their primary importance in solving environmental problems, from quantifying the limits of available resources to the careful technical and economic design of large-scale projects. He wonders what geoscientists can do to change society's low regard for the geosciences, disciplines that he believes should become central in the twenty-first century. Finally, he presents ten proposals to start solving the identity crisis of the geosciences. To the first two points he puts due recognition of the special philosophy of research in the geosciences, which is directed towards the solution of society's problems, and to the need to promote geological education in primary schools, in order to enhance their social dimension.

In April 1991, at a UNESCO workshop in Hamburg which focused on public education, human rights and peace, and aimed to promote a humanistic and intercultural dimension of education, the word 'geoethics' was first used. Kaisa Savolainen, a Finnish woman, then director of the Education Division of UNESCO, used for the first time the word 'geoethics', alongside bioethics, to identify the set of emerging ethical issues related to the development of science and technology, to be integrated into educational programmes at all levels. Savolainen used 'geoethics' without giving a definition but rather as a synonym for environmental ethics, stressing that it 'is an absolute and urgent necessity, since it is a factor that now influences the quality of life for the future and even the survival of humanity' (Savolainen, 1992, p. 45).

In 1992, at the congress of the Geological Society of America, the American geologist Vincent Cronin illustrated the case of a fault (i.e., a fracture in the earth's crust accompanied by displacement of the parts in contact) in the Malibu area in California, with clear indications of seismogenic activity, i.e., suspected of being capable of generating earthquakes (Cronin, 1992). The author pointed out that the formalization of the hazard of this fault by geoscientists would have led to a significant decrease in the economic value of properties located near it, and he questions the ethical responsibilities of the geoscientists involved in that seismic risk assessment. A case like the Malibu fault, which is also quite frequent in other earthquake-prone areas, raises concrete questions of geoethics, since scientific decisions can have political and economic consequences that affect the lives, activities, and property values of local populations.

The repercussions of geological practice on the social fabric are evident, particularly in the field of seismic risk studies, the non-neutral nature of some scientific choices, even if based on scientifically rigorous foundations, and their non-negligible impact on society. Even the simple identification of a fault can raise a series of delicate ethical and social questions: if there is even a small probability that the fault is capable of generating a strong earthquake, how should the population be informed? Who is responsible for communicating the possible danger of that geological object? What should be done in practice to minimise the risks to schools, hospitals, factories and infrastructures? How can we ensure a constructive relationship between geoscientists, decision makers, and citizens when the safety of the population and economic activities are at risk?

The credibility of geoscientists is based on the awareness that their role towards society cannot be neutral since there are numerous geological problems, as well as numerous ethical and social implications to which geoscientists are called to respond. In the last twenty years, the urgency of some environmental issues demands a new scientific commitment by geoscientists, the responsibility to assume their social role as never before. These environmental issues include pollution of air, soil, groundwater and oceans, the search for new energy and mineral sources, the increase in the average temperature of the planet and the need to increase the adaptive response to climate change, the increase in the number of disasters triggered by natural and anthropogenic phenomena, and the reduction of biodiversity.

In a historical moment like the present, in which the whole of society is faced with an ethical crossroads, having to choose between continuing to overexploit the planet's resources or changing the dominant political-economic systems, geosciences are clarifying the deep interconnections and delicate balances of the Earth system, the irreversibility of many phenomena underway, the enormous anthropic impact on natural processes and dynamics, and the limit of available resources. In this scenario, geoscientists can become promoters of a truly collective awareness, leading to the development of a global governance, capable of addressing issues that, beyond political, social, economic and cultural differences, concern the whole of humanity.

For its part, society should be able to grasp the suggestions of geoscientists in order to identify the most suitable political, social, and economic forms for their realistic application, which can ensure environmental, generational, and distributive justice, and overcome unsustainable social and economic paradigms (Blühdorn & Welsh, 2007; Göpel, 2016; Langford, 2016).

For the first time in history, humans are being called upon to respond responsibly to serious global problems, overcoming possible conflicts between different positions and visions, seeking an effective synthesis of scientific results, theories, and models that can lead to a species response that is as shared as possible, to make the transition from ethics to geoethics in the most effective way, for which the need, if not the urgency, is increasingly felt.

References

Battaglia, L. (2012). Alle origini del pensiero ecologico. In L. Battaglia (Ed.), *Un'etica per il mondo vivente—Questioni di bioetica medica, ambientale, animale* (pp. 128–144), Carocci editore, Roma, ISBN 978–8843061648.

Bickford, M. E. (Eds.). (2013). *The impact of the geological sciences on society* (Special Paper 501). The Geological Society of America, ISBN 978–0813725017. https://doi.org/101130/SPE501

Blühdorn, I., & Welsh, I. (2007). Eco-politics beyond the paradigm of sustainability: A conceptual framework and research agenda. *Environmental Politics, 16*(2), 185–205. https://doi.org/10.1080/09644010701211650

Bobrowsky, P., Cronin, V. S., Di Capua, G., Kieffer, S. W., & Peppoloni, S. (2017). The emerging field of geoethics. In L. C. Gundersen (Ed.), *Scientific Integrity and Ethics: With Applications to the Geosciences, Special Publications 73,* 175–212. American Geophysical Union. https://doi.org/10.1002/9781119067825.ch11

BRGM. (2009). *10 core challenges for the geosciences. International year of planet earth special issue* (p. 128). Bureau de Recherches Géologiques et Minières, ISBN 978–2715924789, https://fr.calameo.com/read/0057191219afd5a8faeb3. Accessed 11 November 2021.

Cronin, V. S. (1992). On the seismic activity of the Malibu Coast Fault Zone, and other ethical problems in engineering geoscience. *Geological Society of America, Abstracts with Programs, 24*(7), A284.

Crutzen, P. J. (2002). Geology of mankind. *Nature, 415,* 23. https://doi.org/10.1038/415023a

Franceschelli, O. (2008). *Karl Löwith - Le sfide della modernità tra Dio e nulla* (p. XX+254). Donzelli editore, Roma, ISBN 978–8860362117.

Frodeman, R. (1995). Geological reasoning: Geology as an interpretive and historical science. *Geological Society of America Bulletin, 107*(8), 960–968. https://doi.org/10.1130/0016-7606(1995)107%3c0960:GRGAAI%3e2.3.CO;2

Göpel, M. (2016). The great mindshift: How a new economic paradigm and sustainability transformations go hand in hand. *The Anthropocene: Politik-Economics-Society-Science* (No. 2, p. XXIII+184), Springer Open, ISBN 978–3319437668. https://doi.org/10.1007/978-3-319-43766-8

GSL. (2014). *Geology for society. The geological society of London* (p. 20). https://www.geolsoc.org.uk/geology-for-society. Accessed 11 November 2021.

Ippolito, F. (1968). *La natura e la storia* (p. 147). All'Insegna del Pesce d'Oro, Milano.

Langford, M. (2016). Lost in transformation? The politics of the sustainable development goals. *Ethics & International Affairs, 30*(2), 167–176. https://doi.org/10.1017/S0892679416000058

Leopold, A. (1949). *A Sand County Almanac: And Sketches here and there* (p. 240). Oxford University Press, Enlarged edition (31 December 1968), ISBN 978–0195007770.

Lucchesi, S. (2017). Geosciences at the service of society: The path traced by Antonio Stoppani. *Annals of Geophysics, 60*, Fast Track 7: Geoethics at the heart of all geoscience. https://doi.org/ 10.4401/ag-7413

Moores, E. M. (1997). Geology and culture: A call for action. *GSA Today, 7*(1), 7–11.

Peppoloni, S., & Di Capua, G. (2020). Geoethics as global ethics to face grand challenges for humanity. In G. Di Capua, P. T. Bobrowsky, S. W. Kieffer, & C. Palinkas (Eds.), *Geoethics: Status and Future Perspectives, Geological Society of London, Special Publications, 508*, 13–29. https:// doi.org/10.1144/SP508-2020-146

Peppoloni, S., Bilham, N., & Di Capua, G. (2019). Contemporary geoethics within the geosciences. In M. Bohle (Ed.), *Exploring geoethics—Ethical implications, societal contexts, and professional obligations of the geosciences* (pp. 25–70), Palgrave Pivot. https://doi.org/10.1007/978-3-030-12010-8_2

Reclus, E. (1905–1908). L'Homme et la Terre, Librairie Universelle, Paris, tr. it. in P. L. Errani, a cura di, Elisée Reclus. L'homme. Geografia sociale, Angeli, Milano 1984.

Savolainen, K. (1992). Education and human rights: New priorities. In *Adult education for international understanding, human rights and peace* (pp. 18–19). Report of the Workshop held at Unesco Institute for Educations, April 1991, UIE Reports (No. 11, pp. 43–48).

Wysession, M. E., & Rowan, L. R. (2013). Geoscience serving public policy. *Geological Society of America, Special Papers, 501*, 165–187. https://doi.org/10.1130/2013.2501(07)

Wyss, M., & Peppoloni S. (Eds.) (2015). *Geoethics: Ethical challenges and case studies in earth sciences* (p. 450). Elsevier, ISBN 978–0127999357. https://doi.org/10.1016/C2013-0-09988-4

Zen, E.-A. (1993). The citizen-geologist: GSA presidential adddress. *GSA Today, 3*(1), 2–3.

Chapter 3
From Ethics to Geoethics

3.1 The Problem of Choice

Acting rationally when making decisions means taking into account multiple elements and considerations. To make a choice, the individual must weigh up the various possible options and their consequences, understand which decisions are best and apply them in accordance with the objective to be achieved. In carrying out this logical process, the decision maker takes into account the frame of reference values, hierarchically ordered, which structures their ethical dimension and view of reality, as well as any contingent and occasional issues that may push them towards one choice rather than another.

If we move from an individual dimension to a social group or a human community, up to encompassing the whole of humanity, this decision-making scheme becomes considerably more complicated, since it implies confrontation and dialogue between different stakeholders (from those who will materially make the decisions to those who will put them into practice, up to those who will somehow be affected by them), whose values, objectives, priorities and expectations may differ or even conflict.

Therefore, while individual decisions aimed at obtaining a personal benefit or a result considered useful for the group to which one belongs are relatively easier to make, since the universe of intimate, family, religious and social values that drive the one who acts is unique and personal, when decisions become collective, the lack of common value frameworks can lead to inaction, conflict, extreme positions and incommunicability. If, for example, we consider the issue of climate change or other global problems that require synergic and concerted action by the whole of humanity, the element that immediately emerges and can condition decision-making criteria and intervention methods is the diversity of cultures, lifestyles, visions and value systems that characterise the human mosaic. This diversity cannot block the decision-making process, but must be the starting point for a necessary negotiation between states and within states between the different social groups, aimed at finding a possible and acceptable alignment of values capable of motivating common choices

© The Author(s), under exclusive license
to Springer Nature Switzerland AG 2022
S. Peppoloni and G. Di Capua, *Geoethics*,
https://doi.org/10.1007/978-3-030-98044-3_3

starting from the wealth of points of view. This must include every effort to ensure that choices concerning climate change mitigation and adaptation policies, as well as the commitment to move towards the development of a recycling economy and/or to change production and supply chains in order to reduce their ecological footprint, are translated into actions that are locally defined, i.e., that take into account the cultural, economic, and social context they affect. In addition, since the problems in question are now common to all humanity, the achievement of the targets will have to be structured according to locally agreed and accepted ways and timescales.

For these reasons, the definition of common objectives and priorities on which to base the individual, social, and/or governmental actions of a state is only possible by identifying shared values and ethical criteria capable of guiding towards agreed choices.

Since modern societies perceive the effects of ecological imbalances induced by anthropic action with increasing concern, there is an urgent need to ask what common ethical criteria should guide an interaction with the Earth system to ensure responsible human development. This question opens up a range of crucial issues that question the possibility of finding a sustainable balance between the preservation of life on the planet and the current economic structures of our societies, a balance that is based on the identification of a *safe operating space* for human beings, leading to inclusive and sustainable economic development that does not cross certain ecological thresholds and is capable of ensuring the preservation of the natural system and the protection of human life (Rockström et al., 2009; Steffen et al. 2015). In this context, what is the social role of Earth system experts (geoscientists) and non-experts (ordinary people)?

Geoethics aims to answer these questions and the issues arising from them. Its conceptual substratum and its practical applications are born and rooted in the value that geoscientific knowledge possesses. Even if its development began as a reflection on the role and responsibility of geoscientists and their awareness of being operators at the service of society (Peppoloni & Di Capua, 2012), over time its scope has gone beyond the specific scientific and professional communities, and today its reflections and analyses extend to the whole of society, respecting and incorporating the contribution of all people with different knowledge, skills, experience and perspectives to solve global problems (Mogk, 2020).

Geoethics promotes an analytical and critical attitude based on science and, to do this, it strives to transfer scientific knowledge about the earth's functioning to the various sectors of society so that everyone is able to understand that it is possible to establish a more functional and respectful relationship between human beings and the planet and to build responsible economic, technological, and social development. Geoethics fulfils the vital task of shaping cultural categories and behavioural values based on experience and scientific knowledge (Peppoloni & Di Capua, 2016), capable of outlining a common ethical matrix that responds to the need to find shared operational and decision-making methods, albeit articulated in the complex reality of human cultures.

3.2 From the Definition of Ethics to Geoethics

The theoretical framework, contents, and values underlying geoethics were developed starting from the concept of ethics introduced by the Greek philosopher Aristotle (384–322 BC) in the *Nicomachean Ethics*, which describes ethics as research and reflection on the conduct of human beings and the criteria by which to evaluate behaviour and choices, in order to identify the true good and the means to achieve it. For Aristotle, ethics is an eminently practical science, in which knowledge must be aimed at action. It is therefore also concerned with the moral obligations of human beings towards themselves and others, and what is right when faced with a decision to be made, with the clear objective of striking a balance between good deriving from partisan needs and aspirations and the common good. To simplify, ethics aims to clarify, for a given circumstance, what to do and how to do it, taking into account the consequences of that action. Its function is to guide human beings when they find themselves having to make a choice and make a decision, providing a framework of reference values shared by the social group to which they belong, capable of guiding them towards the good or towards what is most useful and functional for the individual or society at that particular moment.

With reference to the practice of a profession, ethics allows us to identify duties and rights that regulate the professional activity (deontology) of members of a social group, united by the possession of certain technical-scientific knowledge, and methods and tools for their application (Peppoloni & Di Capua, 2018).

In an economically and, to some extent, culturally globalised society, structured in forms of regional and planetary governance, the construction of common value frameworks, albeit often influenced by logics of power and dominance, has created the historical conditions for defining ethics that concern all human beings. If global ethics were invoked in the full development of modern economic globalisation (Küng, 1997), leading to the *Declaration Toward a Global Ethic* (PWR, 1993), similarly the need to address anthropogenic problems in a concerted and effective way will once again require humanity to share ethics towards the Earth system, capable of shaping a new thought and directing the trajectories of cultural, social, and economic development towards more sustainable goals for societies, as suggested by the *Cape Town Statement on Geoethics* (Di Capua et al., 2017).

The ongoing historical process has produced in the mid-twentieth century an increased recognition of universal principles. Codifications such as the *Universal Declaration of Human Rights*[1] establish ethical principles that should guide behaviour, from individuals to states, including dignity, justice, and respect for life. The principle of intergenerational equity is also fundamental to modern societies (Spijkers, 2018). Making rational and responsible choices requires us to apply ethical principles in pursuit of the greater good (Weber, 1919), not only in relation to current societies but also to future societies, since every choice today will produce effects on generations to come (Jonas, 1979). However, the application of principles that we

---

[1] United Nations, *The Universal Declaration of Human Rights*, 1948, http://www.un.org/en/univer sal-declaration-human-rights/. Accessed 11 November 2021.

consider universal may vary in time and space, depending on the social, political, and cultural context. Moreover, although ethics concerns everyone, the duty to act ethically is more stringent for those who hold scientific, political, and social roles of great responsibility and who are entrusted with putting these principles into practice, since the effects of their decisions affect the entire community.

However, this relativism in objectives and priorities for action cannot be reconciled with the close connection existing between the social-ecological systems of the planet and their possible degradation if certain thresholds of systemic sustainability are exceeded. The serious problems caused by Western development models, which have gradually been accepted or imposed at a global level, cannot now be solved by actions limited to local or national contexts but require concerted actions of the implementation methods and coordination at a supranational level. Hence, there is a need to develop global ethics that take into account local experiences and cultures but are also capable of articulating themselves in a general vision aimed at achieving a common goal. If the twentieth century emphasised the value of cultural diversity and the different ethical orientations that characterise each culture, the new millennium cannot fail to be based on those features of unity that are the common denominator of every human culture, in order to build ethics for the earth as a concrete act of the human being who, recalling Élisée Reclus, recognises themselves as 'nature that has become aware of itself'.

Geoethics was born and developed out of this historical necessity, as well as to overcome the fragmentation of environmental ethics in its multiple visions (Marshall, 1993), while drawing from it valuable ideas and reflections for its theoretical development. As suggested in Bohle and Di Capua (2019), geoethics sits at the intersection of environmental ethics (Hourdequin, 2015), sustainability ethics (Oermann & Weinert, 2016) and professional ethics (AGI, 2015; Herkert, 2004), valuing the geosciences and the role of geoscientists from a societal point of view.

Today, the potential of geoethics to propose itself as ethics capable of redefining the relationship between human beings and the Earth system on a global level is more evident than ever.

3.3   The Definition of Geoethics

Geoethics has been defined as 'research and reflection on the values which underpin appropriate behaviours and practices, wherever human activities interact with the Earth system' (Peppoloni & Di Capua 2015, p. 4, 2021; Di Capua & Peppoloni, 2019). In this definition, the subject of reflection is established, and the perimeter of analysis and practice is provided, underlining the need to identify the values on which to base a more responsible and functional interaction between human beings and the planet understood as a system. Furthermore, it states that 'geoethics deals with the ethical, social and cultural implications of geoscience knowledge, education, research, practice, and communication, and with the social role and responsibility of geoscientists in conducting their activities' (Peppoloni & Di Capua, 2015,

p. 4, 2021; Di Capua & Peppoloni, 2019). Therefore, the emphasis is placed on the centrality of geosciences as a *corpus* of technical-scientific knowledge and practices, which allow us to understand the ways in which the Earth system works and can contribute to the construction and dissemination of a scientific culture on which to base responsible management of interactions between human communities and ecological systems. Geoscientists are asked to become more aware of the cultural dimension of geosciences, to recognise the social value of their scientific knowledge, as well as to assume the ethical responsibility of using this knowledge for real progress of society, while respecting the balance between the development of social systems and the protection of ecological systems.

These assumptions are considered fundamental for basic and applied scientific research that looks at actual benefit for the human being, who cannot be considered distinct from the ecosystems in which they are immersed. Without wishing to diminish the value of the intellectual tension and passion required to devote oneself to science, what we wish to emphasise is the necessary commitment to enhancing the social service role of geoscientists, which can give an even deeper meaning to scientific research as a whole, especially if carried out with public funds.

In this perspective, geoscientists are called upon to assume a responsibility that is not only scientific, (i.e., to ensure competence, professionalism, and strict adherence to the scientific methods) but also an ethical responsibility towards society, which arises from the very fact of possessing knowledge that can have important societal repercussions and is therefore no longer merely the intellectual property of the individual but a collective good.

In fact, geoscientists are asked to recognise themselves as moral subjects, to share their knowledge by promoting awareness and responsibility in society, thus contributing to the improvement of human safety and well-being and the preservation of life on the planet.

3.4 The Profound Meaning of Geoethical Thinking

The etymological roots of the word 'geoethics' evoke ancestral concepts concerning the relationship that, as humans, binds us intimately to Earth. Their great modern significance can inspire new paradigms of behaviour towards nature and in the relationships within our societies.

The etymological analysis contributes to the theoretical framework of geoethics, enlightens relevant concepts and provides a deeper understanding of its philosophical matrix, substantiating its meaning as ethics towards Earth.

The prefix *geo* clearly refers to Earth (from the Greek $\gamma \tilde{\eta}$). But its ancient Sumerian base *ga* contains a deeper meaning: 'house, dwelling, place where one dwells' (Semerano, 2007). Therefore, the earth is to be understood more specifically as the place where we live, where our ancestors lived, and where our children should be able to live in safety and health. This is a non-trivial meaning, in which we can rediscover the sense of our relationship with the planet, including the responsibility towards Earth and also for future generations, recalling the fundamental concept of sustainability, that is, the continued use of resources in a long-term perspective (Peppoloni & Di Capua, 2015).

The notion of dwelling can directly refer to the more recent concept of 'human niche' (Bohle & Preiser, 2019; Ellis, 2015; Fuentes, 2016; Low et al., 2019; Meneganzin et al., 2020), to be understood as stated in Fuentes (2016, p. S14): 'contemporary ecological and evolutionary view: it is the dynamic N-dimensional space in which an organism exists - the totality of the biotic and abiotic factors that make up an organism's main context for evolutionary dynamic'. The 'human niche' summarises the notion that 'humans construct ecological, technical, and cultural niches that influence the structure of evolutionary landscapes' (Fuentes, 2016). Earth, understood as the human niche, is the physical but also cultural and technological place that current generations have an ethical duty to best preserve and transfer to future generations.

The etymological analysis of the word *ethics* leads to a double meaning. *Ethics* derives from the Greek ἔθος, which means 'custom, habit'. This term has the same origin as εἴωθα, a verbal form of the Greek perfect, meaning 'I have custom, I have familiarity' (Liddell & Scott, 1996). A sense of belonging to a community, be it a family or a wider social group, is inherent in the words 'familiarity' and 'custom' but what determines familiarity and, thus, the habit of a behaviour? There is a trace of this passage in the Semitic root *edum*, which means 'experience, to be expert in' (Semerano, 2007). In other words, one experiences an event, a circumstance, acquires knowledge of it, becomes familiar with it, becomes an expert in it. From this moment on, the acquired experience will enable the individual to choose the most appropriate behaviour or habit for a given circumstance or event. But, if the word ethics is related to ἦθος, it also refers to the personal characteristics or habits of the individual (Liddell & Scott, 1996). Both nouns derive from the same root *sweth* (comparable to the Latin verb *suesco*: Ernout & Meillet, 1994), highlighting the dual nature of human beings, who are both individuals and members of a community. Thus, *ethics* has a double meaning, one referable to the social sphere, the other to the individual sphere.

The same conclusion can be reached by going back to the Akkadian language. Starting from the Akkadian base *esdu*, ethics is given the meaning of 'foundation, social regulations', and by extension also the meaning of 'assurance of continuity' (Semerano, 2007), therefore, once again, being a reference to the social dimension. If we consider the Akkadian basis *betu*, ethics means 'home, dwelling, refuge', with the reference to a more intimate and personal space in every human being. Finally, in relation to the Akkadian base *ettu*, the word *ethics* is charged with the value of 'character, distinctive mark of an individual, characteristic trait of a person', once again a reference to the individual sphere (Semerano, 2007).

To summarise, it seems that the word ethics can be associated with a dual meaning, whereby it contains a sense of belonging and conformity to a social dimension, and it also relates to individual behaviour. Ethics concerns both the social sphere, regulating interactions between individuals belonging to a community or social organisation, and the personal sphere, defining what distinguishes an individual, outlining their adherence to an inner dictate. Ethics, therefore, means conforming to something that includes us and, at the same time, conforming to one's own individual nature. These two existential conditions (social and individual) coexist in the word ethics, indicating acting both for the personal and the common good. As Semerano (2007, p. V) states, 'Words, more enduring than any metal, are authoritative witnesses to our ancient unity', to which one might add that by rediscovering their profound meaning one creates the foundations for a future unity among people.

These reflections can be extended to geoethics, to give substance to its definition. On one hand, it is the investigation of and reflection on the behaviours of the human agent towards the society of which they are a part and the Earth system that hosts them; on the other hand, it is the analysis of the relationship between the human agent and their own individual actions.

For this reason, the human agent is called not only to an individual responsibility but also to a social and environmental responsibility, as they are intimately linked.

3.5 The Fundamental Characteristics of Geoethics

Geoethics is increasingly recognised as an emerging field within the geosciences (Bobrowsky et al., 2017). In recent years, a number of geoscientists have been engaged in defining the conceptual framework of geoethics, progressively outlining its definitions, contents, methods, tools, main topics of interest and an increasingly shared vision. Through this participatory process, geoethics now has consolidated theoretical foundations that allow its practical application, not only within the different disciplinary fields of geosciences but also in society (Peppoloni & Di Capua, 2017, 2020).

A distinctive feature of geoethics is that it is centred on the human agent. The individual is at the centre of an ethical reference system in which virtuous behaviour (based on dignity, respect, honesty and prudence) contributes to establishing a relationship between human beings and the Earth system that is functional and respectful of all the elements that make up the ecological systems, both biotic and abiotic, the latter being recognized as the essential substratum of natural reality. The geoscientist would add, to the list of virtuous behaviour, to provide correct and competent scientific information.

Geoethics is characterised as ethics of virtue with reference to the human agent, based on the principles of dignity, freedom, responsibility, awareness, justice and respect (Peppoloni & Di Capua, 2020) and implemented through a participatory process, aimed at resolving issues starting from the scientific knowledge of the planet.

Geoethics emphasises individual responsible action, based on the adoption of social and, in the case of geoscientists, professional reference values. Its development and application take place within a pragmatic process of open and continuous review of choices, supported by the work of scientists aware of their social role in the service of society.

Responsibility is the bridge that binds the cultivation of virtues in the development of solid moral character on the part of the individual and his/her actions, in the vast context of the relationships in which he/she is immersed. In this sense, geoethics is ethics of virtue but also ethics of responsibility, since it implies conscious acceptance of the consequences of one's actions in a system of complex relationships, which will inevitably be modified by one's choices.

Behaviour inspired by geoethics is based on scientifically founded choices. Geoscientists, with their wealth of knowledge on the dynamics of the planet, are able to suggest choices, behaviours, practices and functional ways of interacting with the Earth system, even and especially to those who are not geoscientists, such as ordinary citizens, policy makers or legislators.

Choices in line with the geoethical vision cannot be affected by radicalism but must take into account the variety of cultures and human social conditions present globally. As well, they cannot ignore the physical–chemical-biological peculiarities of the territories subject to anthropogenic interventions. These choices must therefore be spatially and temporally contextualised. Any analytical and critical approach to environmental problems cannot ignore justice, equity and inclusiveness if it is to guarantee to the various human groups and, by extension, to future generations, the same opportunities for social, economic, and cultural development in a natural environment that is not degraded, even from an aesthetic point of view. It is evident that any choice or solution affecting the territory that does not consider local realities risks being perceived as an imposition and may provoke contrary, even violent, reactions of the population, which may have different points of view, beliefs, needs, objectives and expectations. This is the condition that often arises when decisions are not the result of an inclusive and participatory process with citizens, particularly if they concern projects with a major socio-environmental impact such as infrastructure networks, underground drilling, the construction of reservoirs, landfill sites and nuclear waste deposits.

Similarly, choices that do not take into account the possible progressiveness of environmental impacts over time may appear to be effective solutions in the short term but lead to additional problems over longer periods or vice versa.

The apparent relativism of action should not be understood as an intrinsic risk within geoethics. On the contrary, the definition of prescriptive behaviours and norms that underestimate the importance of the context in which they are to be applied could have the opposite effect, inducing antagonistic tensions and an 'a priori' refusal by the population involved. The geoethical approach takes into account the spatio-temporal complexity of existing physical and social realities, identifying their technical, environmental, economic, cultural and political limits, being aware that similar problems may require different solutions, geoethically correct in different contexts.

In the geoethics view, the human agent is guided by a set of individual, professional, social and environmental values which support their responsibilities. Those responsibilities are divided into four levels of interaction, corresponding to as many consecutively larger and more complex domains of human experience: the self, the social group(s) to which one belongs (including the professional groups), society, and the environment. In these four domains the individual acts consciously, according to an analytical and prudent approach based on the principle of responsibility. A genuinely geoethical decision can only arise from a responsible choice, whose fundamental prerequisite is the freedom to choose (Peppoloni et al., 2019). Freedom is an indispensable condition for making an ethical choice, which is aware and reasoned and not forced or imposed, resulting from deep examination of all the possible consequences, including negative consequences that may occur.

The essential characteristics outlined above foreshadow the innovative potential of geoethics and the importance of proposing it as a transnational and cross-cultural moral reference.

References

AGI. (2015). *Guidelines for ethical professional conduct.* American Geosciences Institute. https://www.americangeosciences.org/community/agi-guidelines-ethical-professional-con duct. Accessed 29 March 2022.

Bobrowsky, P., Cronin, V. S., Di Capua, G., Kieffer, S. W., & Peppoloni, S. (2017). The emerging field of geoethics. In L. C. Gundersen (Ed.), *Scientific Integrity and Ethics: With Applications to the Geosciences, Special Publications, 73*, 175–212. American Geophysical Union. https://doi.org/10.1002/9781119067825.ch11

Bohle, M., & Di Capua, G. (2019). Setting the scene. In M. Bohle (Ed.), *Exploring geoethics—Ethical implications, societal contexts, and professional obligations of the geosciences* (pp. 1–24), Palgrave Pivot. https://doi.org/10.1007/978-3-030-12010-8_1

Bohle, M., & Preiser, R. (2019). Exploring societal intersections of geoethical thinking. In M. Bohle (Ed.), *Exploring geoethics—Ethical implications, societal contexts, and professional obligations of the geosciences* (pp. 71–136), Palgrave Pivot. https://doi.org/10.1007/978-3-030-12010-8_3

Di Capua, G., Peppoloni, S., & Bobrowsky, P. (2017). The Cape Town statement on geoethics. *Annals of Geophysics, 60*, Fast Track 7: Geoethics at the heart of all geoscience. https://doi.org/10.4401/ag-7553

Di Capua, G., & Peppoloni, S. (2019). *Defining geoethics.* Website of the IAPG—International Association for Promoting Geoethics, http://www.geoethics.org/definition. Accessed 11 November 2021.

Ellis, E. C. (2015). Ecology in an Anthropogenic Biosphere. *Ecological Monographs, 85*(3), 287–331. https://doi.org/10.1890/14-2274.1

Ernout, A., & Meillet, A. (1994). *Dictionnaire étymologique de la langue Latine (retirage de la quatrième édition)* (p. 820). Éditions Klincksieck.

Fuentes, A. (2016). The extended evolutionary synthesis, ethnography, and the human Niche: Toward an integrated anthropology. *Current Anthropology, 57*, S13–S26. https://doi.org/10.1086/685684

Herkert, J. R. (2004). Microethics, macroethics, and professional engineering societies. *Emerging technologies and ethical issues in engineering* (pp. 107–114) (Papers from a Workshop 2004). National Academy of Engineering, National Academies Press, ISBN 978–0309092715.

Hourdequin, M. (2015). *Environmental ethics—From theory to practice* (p. 256). Bloomsbury Academic, ISBN 978–1472510983.

Jonas, H. (1979). Das Prinzip Verantwortung: Versuch einer Ethik für die technologische Zivilisation. Suhrkamp, Frankfurt/M. *The imperative of responsibility: In search of ethics for the technological age* (translation of Das Prinzip Verantwortung) trans. Hans Jonas and David Herr (1979). ISBN 0–226–40597–4 (University of Chicago Press, 1984), ISBN 0–226–40596–6.

Küng, H. (1997). A global ethic in an age of globalization. *Business Ethics Quarterly, 7*(3), 17–32. https://doi.org/10.2307/3857310

Liddell, H. G., & Scott, R. (1996). *A Greek-English Lexicon* (with a Revised Supplement) (p. 2448). Clarendon Press, ISBN 978–0198642268.

Low, F. M., Gluckman, P. D., & Hanson, M. A. (2019). Niche modification, human cultural evolution and the Anthropocene. *Trends in Ecology & Evolution, 34*(10), 883–884. https://doi.org/10.1016/j.tree.2019.07.005

Marshall, A. (1993). Ethics and the extraterrestrial environment. *Journal of Applied Philosophy, 10*(2), 227–236.

Meneganzin, A., Pievani, T., & Caserini, S. (2020). Anthropogenic climate change as a monumental niche construction process: Background and philosophical aspects. *Biology & Philosophy, 35*, 38. https://doi.org/10.1007/s10539-020-09754-2

Mogk, D. W. (2020). The intersection of geoethics and diversity in the geosciences. In G. Di Capua, P. T, Bobrowsky, S. W. Kieffer & C. Palinkas (Eds.), Geoethics: Status and future perspectives. *Geological Society of London, Special Publications, 508*, 67–99. https://doi.org/10.1144/SP508-2020-66

Oermann, N. O., & Weinert, A. (2016). Sustainability ethics. In H. Heinrichs, P. Martens, G. Michelsen, & A. Wiek (Eds.), *Sustainability science* (pp. 175–192). Springer, Dordrecht. https://doi.org/10.1007/978-94-017-7242-6_15

Peppoloni, S., Bilham, N., & Di Capua, G. (2019). Contemporary geoethics within the geosciences. In M. Bohle (Ed.), *Exploring geoethics—Ethical implications, societal contexts, and professional obligations of the geosciences* (pp. 25–70). Palgrave Pivot. https://doi.org/10.1007/978-3-030-12010-8_2

Peppoloni, S., & Di Capua, G. (2012). Geoethics and geological culture: Awareness, responsibility and challenges. *Annals of Geophysics, 55*(3), 335–341. https://doi.org/10.4401/ag-6099

Peppoloni, S., & Di Capua, G. (2015). The meaning of geoethics. In M. Wyss & S. Peppoloni (Eds.), *Geoethics: Ethical challenges and case studies in earth sciences* (pp. 3–14). Elsevier. https://doi.org/10.1016/B978-0-12-799935-7.00001-0

Peppoloni, S., & Di Capua, G. (2016). Geoethics: Ethical, social, and cultural values in geosciences research, practice, and education. In G. R. Wessel & J. K. Greenberg (Eds.), Geoscience for the public good and global development: Toward a sustainable future. *Geological Society of America, Special Paper, 520*, 17–21. https://doi.org/10.1130/2016.2520(03).

Peppoloni, S., & Di Capua, G. (2017). Geoethics: Ethical, social and cultural implications in geosciences. *Annals of Geophysics, 60*, Fast Track 7: Geoethics at the heart of all geoscience. https://doi.org/10.4401/ag-7473

Peppoloni, S., & Di Capua, G. (2018). Ethics. In P. T. Bobrowsky, & B. Marker (Eds), *Encyclopedia of engineering geology*, Encyclopedia of Earth Sciences Series, Springer. https://doi.org/10.1007/978-3-319-12127-7_115-1

Peppoloni, S., & Di Capua, G. (2020). Geoethics as global ethics to face grand challenges for humanity. In G. Di Capua, P. T. Bobrowsky, S. W. Kieffer, C. Palinkas (Eds.), *Geoethics: Status and Future Perspectives, Geological Society of London, Special Publications, 508* 13–29. https://doi.org/10.1144/SP508-2020-146

Peppoloni, S., & Di Capua, G. (2021). Current definition and vision of geoethics. In M. Bohle, & E. Marone (Eds.), *Geo-societal narratives—Contextualising geosciences* (pp. 17–28). Palgrave Macmillan. https://doi.org/10.1007/978-3-030-79028-8_2

PWR. (1993). *Declaration toward a global ethic.* Council for a Parliament of the World's Religions. https://parliamentofreligions.org/sites/default/files/docs/global_ethic_pdf_-_2020_update.pdf. Accessed 29 March 2022.

Rockström, J., Steffen, W., Noone, K., Persson, A., Chapin, F. S., Lambin, E. F., Lenton, T. M., Scheffer, M., Folke, C., Schellnhuber, H. J., Nykvist, B., de Wit, C. A., Hughes, T., van der Leeuw, S., Rodhe, H., Sörlin, S., Snyder, P. K., Costanza, R., Svedin, U., Falkenmark, M., Karlberg, L., Corell, R. W., Fabry, V. J., Hansen, J., Walker, B., Liverman, D., Richardson, R., Crutzen, P., & Foley, J. A. (2009). A safe operating space for humanity. *Nature, 461*(7263), 472–475. https://doi.org/10.1038/461472a

Semerano, G. M. (2007). Le *Origini della Cultura Europea: Dizionari Etimologici* (p. C+726). Olschki Editore, Firenze, ISBN 978–8822242334.

Spijkers, O. (2018). Intergenerational equity and the sustainable development goals. *Sustainability, 10*(3836). doi:https://doi.org/10.3390/su10113836

Steffen, W., Richardson, K., Rockström, J., Cornellingo, S. E., Bennettreinette, Fetzer, I., Bennett, E. M., Biggs, R., Carpenter, S. R., de Vries, W., de Wit, C. A., Folke, C., Gerten, D., Heinke, J., Mace, G. M.,Persson, L. M., Ramanathan, V., Reyers, B., & Sörlin, S. (2015). Planetary boundaries: Guiding human development on a changing planet. *Science, 347*(6223), 1259855–1259855. https://doi.org/10.1126/science.1259855

Weber, M. (1919). *Politik als Beruf—Gesinnungsethik vs. Verantwortungsethik.* Translation in English: https://www.academia.edu/26954620/Politics_as_Vocation.pdf. Accessed 29 March 2022.

Chapter 4
The Concept of Responsibility

4.1 Geoethics and Responsibility

Geoethics has developed within the scientific community to investigate the meaning and value of the geosciences, analysing their rational categories, possible perspectives, uncertainties and cognitive limits, with the aim of understanding how their modes of action conform to a specific vision of reality and how this vision, in turn, can modify the interaction of human action with natural reality (Doglioni & Peppoloni, 2016).

The aim of geoethics is to identify shared values on which to base strategies and operational procedures that are compatible with respect for natural forms and processes, the vocation of the territory, and the health and safety of human communities, strategies that are scientifically defined and contextualised in time and space, i.e., those that take into account different temporal perspectives in the analysis of expected or possible effects and the diversity of territorial, social, cultural, political and economic contexts existing in different places (Peppoloni et al., 2019).

Geoethics is a tool for raising the scientific community's awareness of its deontological, ethical, social and environmental duties and, thanks to its recent theoretical structuring, it is now ready to be proposed to society as a whole as a way of understanding human life on the planet, in the light of an ethical, social, and cultural reference framework capable of directing human actions towards socially sustainable and eco-compatible forms of coexistence (Peppoloni & Di Capua, 2020).

In the scientific and technical sphere, geoethics places the individual geoscientist at the centre of ethical action, calling on them to behave responsibly in the exercise of their profession and the social role they are required to play. In a sphere extended to all human beings, in line with Stoppani's thought, geoethics recognises the human agent as a 'geological force' capable of moulding the Earth system and, in virtue of this prerogative, assigns them an ethical responsibility arising from their awareness of being a modifier of ecosystems. Geoethics considers human beings as the natural conscience of the planet (Peppoloni & Di Capua, 2017). Only by recognising themselves as a 'geological force' in action can they become aware and take responsibility for working for their own good and for the protection of social-ecological systems. This responsibility should be understood as a synthesis of individual, social, environmental, cultural and aesthetic experience, aimed at also guaranteeing acceptable living conditions on Earth for non-human living entities and at adopting conservation actions compatible with their own vital needs, recognising their value. Any modification of the earth's environments will therefore only be possible, if necessary, through behaviour that is aware of the consequences and is as respectful as possible of geodiversity and biodiversity.

4.2 The Meaning of Responsible Action

The concept of responsibility is central to geoethics (Peppoloni & Di Capua 2015, 2017, 2020, 2021; Peppoloni et al., 2019) in its literal meaning. The word 'responsibility' (from the Latin *respondere*) expresses the commitment of the individual agent to answer to someone for their actions and related consequences, provided that they are acting in a condition of freedom of choice. Responsibility includes the duty to satisfactorily perform a task, which in case of error may entail 'guilt' and thus a 'sanction', a 'penalty'. In some cases, this is to be understood as an economic or legal penalty; this is the case, for example, of a negligent geologist who makes the wrong calculations to stabilise a slope and, consequently, a landslide occurs with damages and victims. But, in other cases the penalty to which a geoscience expert is subjected is the very loss of professional credibility (both as individual and as a socially recognised professional category) that an irresponsible action can entail, the failure of their technical and scientific role, for not having been able to support the community in facing and solving a geological problem and, therefore, in having failed in the very mission/function of geoscientist (Peppoloni & Di Capua, 2017, 2018).

Beyond the legal considerations on guilt, in general terms responsibility can involve limitation of freedom of choice or personal action, since it implies having to take into account the other, in particular the consequences that others may suffer as a result of the agent's decisions. The other conditions these decisions and can even become a 'judge' or 'evaluator' of their justness; therefore, a responsibility in which 'the two freedoms, of the self and of the other, are at stake, which must find the way and measure of coexisting in the balance between autonomy and limitation. Responsibility arises when one becomes aware of letting also be the freedom of the other' (Franco, 2015), and this presupposes the simultaneous awareness of the autonomy of the individual and of the network of relationships in which they are immersed, which limit that autonomy.

Therefore, responsibility implies an attribution of value to the one (the individual) or to those (a group of individuals) or to that (a non-human living being, an object, a technological system, a natural environment or, more generally, the whole Earth system) that directly or potentially suffer the consequences of the choices made by an individual agent or a group of individuals, and the ability to assess and/or judge and/or react to the rightness or wrongness, acceptability or unacceptability, usefulness or futility of one choice over another.

If an individual choice has consequences only for the person making it, the goodness or otherwise of that choice will be assessed by the subject considering the full, partial or unsatisfactory achievement of the objective they set themselves. In this case, the agent is answerable for their choices and actions only to themselves. If, on the other hand, the choice of an individual or a group has consequences for a social community, the latter will be able to assess the consequences according to its own reference values and/or the tangible effects that will be manifested. The individual or group acting will be morally and/or legally answerable to the community that has suffered the effects of its action. Finally, if the human choice has consequences for the physical-chemical-biological environment, then it will have to be responsibly confronted with the reaction of the ecosystem to that anthropic action, which is configured as a 'judgement of nature'. Nature will modify its evolutionary trajectory, with consequences that may be resolved in a reduction of complexity and diversity, or even in a restriction of the space of habitability for human beings and other living forms, or in the degradation of non-living elements.

If our decisions and consequent actions do not take into account the processes, dynamics, and complex interactions of nature, it will be nature itself that will have to find new balances through evolutionary adaptations that may also have negative consequences for human beings. From this point of view, a pandemic in the twenty-first century can be considered an inevitable natural response/reaction to the irresponsible increase in human pressure on ecosystems, which can only be remedied by modern scientific research in the medical field.

Therefore, responsibly managing the interaction between human beings and Earth means being accountable for the often unpredictable anthropogenic effects on the planet's ecosystems, which can lead to a range of negative consequences for the safety and health of humans and other life forms. This ecological responsibility belongs both to the individual and to the community in its constituent groups, or to society as a whole.

In this perspective, geoethics is ethics of responsibility that cannot be limited to guiding choices only within a restricted dimension of human experience but is a common substratum on which to set up a global praxis, coherent and science-oriented, that assumes the commitment to responsible human development, enlightening the conscience of the individual in relation to themselves and to all the social and environmental relationships in which they are immersed. However, this responsibility cannot arise, as for Jonas (1979), only from humans' fear of being faced with the risk of their extinction due to unawareness of their limits, nor can it be based on apocalyptic prophecies. Instead, geoethics arises from a humanity renewed in the awareness of its ethical dimension, its decision-making capacity and the intrinsic value of any living or non-living element with which it interacts on the planet.

4.3   The Four Geoethical Domains of Human Experience

The human agent, with their universe of reference values, is at the centre of a network of complex relationships, free or forced, which are structured on the basis of contingencies, needs, opportunities, interests, traditions and beliefs. Daily choices are formed and articulated within this framework of interactions, of which the relationship with oneself represents the deep root, and on which the verification of one's identity is based and the relational system of one's own experiences is structured. Geoethics considers four dimensions (or geoethical domains) of human existential experience to be fundamental: the relationship with the self, the more direct interpersonal relationships, the role and relationships within the larger social community, and the interaction with the natural environment and the Earth system in general, which can be defined, in short, as the individual, interpersonal, social and environmental dimensions.

According to the geoethical view, the human agent examines their choices in relation to these domains of experience, which move from the individual level to consecutively larger, more complex and intricate levels (Bohle & Di Capua, 2019; Mogk et al., 2017; Peppoloni et al., 2019). The consistency of behaviours across the four types of relationships indicates that adherence to the geoethical value system is genuine, the result of a conscious and gradual awareness and not superficial or opportunistic. In fact, respect and understanding for the other is not possible when there is no clarity towards oneself. Even if the promotion of environmentalist messages in society may have an absolute value in its aims, if it is not transformed into concrete actions on the level of personal behaviour and in relation to the groups with which one enters into relations, it ends up assuming an ambiguous character which others cannot grasp as authentic; in other words, adhering to the values proposed. As a result, those messages will remain ineffective and inconsistent, and the individual who proposes them will be perceived as having little credibility.

From the geoethical perspective, the highest responsibility for the human agent is to freely and consciously adhere to a system of values that they will then, for consistency and authentic acceptance, have to take as a reference for their actions in all the four ethical domains of their experience, from the individual dimension up to encompassing the whole Earth system.

4.4   Responsibilities of Geoscientists and the Human Agent

As already indicated, geoethics initially developed as an internal reflection within the geoscience community, as professional ethics that progressively expanded to the complexity of social relations (Bohle & Di Capua, 2019; Peppoloni & Di Capua, 2020; Peppoloni et al., 2019). Its theoretical framework has been built on analysis of the identity of geoscientists, their relationships and the role they play in the spheres in which they make choices and act, from the academic to the research or teaching sector, from the professional to the industry field. However, over time the key concepts of geoethics have been presented outside the world of geosciences and are becoming a valuable reference for society as a whole. Geoethics offers analytical, critical, and scientifically grounded attitudes towards issues concerning the interaction between human beings and nature, defining cultural categories and behavioural values based on experience and scientific knowledge (Peppoloni & Di Capua, 2016, 2020, 2021), which help to orient towards more responsible individual and social choices. Transferring such an attitude to society, understood in all its components (from political decision-makers to legislators, technicians, operators of mass media, and citizens) means promoting responsible economic, technological, and social development, based on well-considered political decisions and their short- and long-term consequences.

4.4.1 Individual Responsibility

The first responsibility of the geoscientist is towards oneself, in conducting one's work to the best of one's ability (Abbott, 2017; Andrews, 2014; Mayer, 2015; Mogk, 2017; Peppoloni et al., 2015). This means pursuing excellence in science and technology, adopting appropriate methods and technologies in theoretical and applied scientific research and best professional practices, being aware of the limits of one's knowledge and skills, and acting accordingly within these limits.

First of all, it is essential to always check the reliability of the sources from which one draws the information and data used in one's research, while the results produced and interpretations must be reported in a complete and objective way, without alterations aimed at strengthening one's hypotheses and theories. A responsible geoscientist is committed to ensuring their continuous professional development and improvement of their scientific knowledge throughout the course of the career to ensure adequate and continuous technical and scientific competence. Furthermore, responsible geoscientists assess and report the degree of uncertainty in their hypotheses and data and clarify any gaps in their observations. Finally, they avoid finding themselves in conflicts of interest that would undermine their freedom of choice, weighing up pressure, intimidation, and other undue interference in their activities that could compromise their ability to act ethically. And, so they are not caught unprepared or accused later, they do not neglect to declare potential conflicts of interest that may arise in the future, even if hindering actions do not exist immediately.

Respect for one's own competence is reflected in the responsibility to recognise the superior competence of others, particularly in those scientific fields in which one does not have solid training, skill, and experience. After all, there are many disciplines within geosciences, and it is impossible to have in-depth knowledge in every area without adequate training. Similarly, one cannot consider oneself capable of challenging theories or scientific results in disciplinary fields in which other colleagues have proven expertise. If these simple rules of conduct were observed, one would not fall into the specious and misleading controversies that often agitate the scientific community and that, once they cross the boundaries of peer-to-peer discussion and enter the public debate, can go so far as to threaten the very understanding of phenomena by non-experts or manipulate arguments for political purposes, as in the case of the delicate issue of climate change or discussions on the SARS-CoV-2 pandemic.

Like other scientists, geoscientists cannot act outside the rules of science, denying its epistemological foundations. Therefore, they cannot base their statements on unverifiable hypotheses or unreliable data, or conduct their research without adopting the highest standards of the scientific methods, those observational, analytical, critical and experimental procedures that allow the best application of results.

Science is a powerful tool in the hands of human beings. Each individual constantly interacts with the Earth system, both directly, by being clearly aware to be a modifier of social-ecological systems, and indirectly, as an end-user in the supply chains of products, goods, and services. One can negatively affect the environment either by using a car that emits particulate matter and greenhouse gases, or by purchasing consumer goods based on overexploitation of non-renewable natural resources. The individual has a responsibility to keep themselves informed, a duty to know the reality in which they are immersed, since personal choices and the resulting actions have inevitable consequences on natural systems. The State, in turn, must guarantee the right to correct information and education, because it is through knowledge that awareness is generated and, in turn, a sense of responsibility towards the common home, of which the territory with its ecosystems is an inalienable heritage.

It is possible to protect one's health and safety by adopting behaviours that contribute, within the limits of personal possibilities, to reducing pollution, protecting against natural risks, using resources prudently, protecting and enhancing the territory, including its aesthetics. Abandoning waste in the environment, building a house in a flood prone zone or on the slope of a volcano, disregarding anti-seismic regulations, as well as venturing into a underpass during a flood, or not giving up off-piste skiing despite the avalanche warnings of the authorities, are not mere naivety, but behaviours to be avoided, whether done in good or bad faith, because of the tragic consequences they may entail.

Nowadays there is the possibility of accessing high-quality scientific information, albeit with the necessary limitations, such as that provided by research organizations and institutes, universities and the civil protection departments. Relying on unreliable Internet sites to improve one's own knowledge on risks is an act of irresponsibility, first and foremost towards oneself, whereas adequate knowledge of the hazardous situations that could affect us allows to not be caught unprepared and, if necessary, to act, combining preparation with common sense.

Understanding this means also being clear about one's social function within the reference social group. As citizens, it is our responsibility to take an active part in the management of a safer, more efficient, and sustainable society, and not to completely delegate personal safety and protection to the government, with consequent loss of one's freedom of action.

The responsible human agent has respect for individual dignity, is consistent with their own inner human nature and qualities, and is committed to challenge automatisms and rigid positions in order to prepare the ground for the critical analysis of reality and the welcome of change, in the rigour of a scientific knowledge that does not allow shortcuts.

4.4.2 Interpersonal Responsibility

Modern research and practice in the field of geosciences are characterised by an increasing multidisciplinary cooperation, which is necessary to grasp the structural and dynamic complexity of ecological systems and to offer global solutions, taking into account the different variables involved. Geoscientists working in the research field conduct their activities in scientific teams composed of different specialists with complementary competencies. Even in the professional field, professionals deal with the complexity of geo-environmental issues, such as those related to large-scale infrastructures, in cooperation with other experts. For example, in seismic hazard studies geologists, physicists, mathematicians, computer scientists and historians work together to define the parameters that engineers then use to measure the structures that must withstand a given earthquake. Similarly, geologists, geomorphologists, hydraulic and structural engineers, hydrogeologists, biologists and geotechnical engineers work together on the construction of a dam, and only their cooperation is capable of framing the infrastructure to be built not only in the objective complexity of its design but also in the vast geo-bio-ecological system that will host it, taking into account all the inevitable direct and indirect impacts associated with it. The same applies to studies dealing with climate, natural and anthropogenic hazards, pollution or sea dynamics. The importance of the ways in which relationships are established within research and professional groups has grown and increased the responsibilities that the individual geoscientist has to assume towards their colleagues.

Interpersonal relations in one's own working environment must be handled with great care. Among collaborators there is the responsibility to relate to colleagues in an honest and polite manner, respecting each other's ideas, accepting debate without preclusion and welcoming each other's views, competencies, and methods, as long as they meet the best available technical and scientific standards. Cooperation between individuals fosters personal intellectual and professional development without discrimination, and the competition that can sometimes arise between colleagues cannot justify disloyalty and meanness but, on the contrary, should lead to giving one's best. The sharing of information and data is a fundamental step towards optimising available resources, avoiding duplication and overlapping, and is an indispensable condition for achieving the best result in the shortest possible time. In this context, respect for intellectual property is the key principle on which recognition of the merits of colleagues and respect for their scientific and professional value are founded. Similarly, recognising that a colleague is better prepared to carry out a particular task than we are is an act of loyalty, the lack of which is not only a gesture of professional misconduct but a threat to the quality of scientific activities and serious misconduct towards those who are the end users of those activities.

In addition to their relationships with colleagues, the geoscientist is immersed in a network of relationships (family, professional, social) that involve listening, dialogue, and negotiation of individual choices. Others can be a limit to one's possibilities and, at the same time, a resource, provided that the relationships established allow for an effective and useful plurality of visions. However, even those who are not geoscientists have their own responsibility in the interpersonal relationships existing within their own network, which can influence their decisions. That responsibility calls for an open and inclusive attitude that respects the dignity, opinions, fundamental rights of the other, and ensures equal treatment and opportunities regardless of diversity. Any form of ethnic, religious, cultural, social, political, and economic or gender discrimination also stems from the inability to recognise the potential of the other and understand the advantage of putting it at the service of the group or wider community of belonging.

For example, discrimination against women deprives society of the intelligence, sensitivity, knowledge, experience, and wealth of ways of seeing reality that women can offer, denying the possibility of society to make use of an important contribution to problem solving (Eastin, 2018). On the website of the United Nations dedicated to climate change, one can read that it is women who face the greatest risks and burdens from the impacts of climate change, especially in situations of poverty. In fact, most of the world's poorest people are women and women have little access to decision-making processes and employment market. This exacerbates inequalities and prevents women from fully contributing to the planning, design and implementation of climate-related policies. Instead, women could play a key role in responding to climate change, as they know the local reality better than anyone, often having to deal with sustainable resource management at household and community levels. The United Nations also points out that the political involvement of women would ensure greater attention to the needs of citizens, as well as more effective cooperation between ethnic groups, ensuring peace and social well-being. In climate-related projects and policies developed at the local level, better results were found when women were in charge of leadership, while their exclusion led to a decrease in policy effectiveness and an increase in existing inequalities.

The other is an opportunity to improve one's knowledge, skills, ability to analyse and respond to problems, while respecting the principle of reciprocity which must guarantee fairness and justice in relationships, so that forms of prevarication and imposition are not generated. Dialogue between parties is the fundamental tool for making relationships solid, lasting, and functional and for finding an acceptable balance between different positions. The responsible human agent is aware that, while their individual behaviour may have some impact on reality, it is precisely cooperation, in a climate of genuine co-participation in decisions and objectives that can increase the effectiveness of positive actions by individuals. Environmental associations, which propose changes in our individual behaviours to improve the conditions of the natural environment and our interaction with the planet, is an expression of the desire to find a synergy between individuals, recognising themselves in a matrix of values that can unite and guide towards common goals. Overcoming natural differences of opinion through dialogue can make all the difference when it comes to reaching agreement and making ecologically oriented decisions.

Lack of responsibility in interpersonal relationships can lead to increased vulnerability of human communities, whatever change should occur. From defence against natural risks to adaptation to climate change, all processes and policies that affect socio-ecological systems benefit from the existence of networks of interpersonal relationships that create solidarity and cooperation among individuals in sharing common health and safety goals.

4.4.3 Social Responsibility

The geoscientist produces knowledge and designs solutions for the direct and indirect benefit of society. The responsibility to serve society in the most effective way possible takes the form of supporting its overall development and ensuring its security through the use of methods, techniques, and tools developed to understand and solve geo-environmental problems, and contributing to structuring the decision-making process that leads to the reduction of impacts on social-ecological systems through the implementation of effective choices.

The area of use of geoscience knowledge and the sphere of action condition the forms of interaction of individual experts with society. A geoscientist working in industry also has stringent obligations and responsibilities in relation to the objectives of the company for which they work. However, corporate interest, often also linked to profit, prevails over public interest and the resulting choices may conflict with the collective good. On the contrary, for those who carry out their activities in the academic or public research sphere, it is easier to give priority to the collective interest (or at least it should be) and, therefore, to identify with their social role.

As an example, geoscientists working in the mining or oil sector, where strong negative impacts on the environment and local communities are at stake, often find themselves living the contradiction between their universe of ethical and social values and corporate expectations, which demand specific operational choices. Therefore, it can happen that under corporate pressure they are forced to change their technical choices in order to comply with corporate policies, even if they do not agree with them, finding themselves in a position of subordination that does not allow them to act ethically. Their decisions will be in antithesis with what professional competence and their own conscience would indicate as sensible and necessary actions. If this is not the case, they could carry out their work to the best of their abilities, acting in accordance with the law, cooperating constructively with their colleagues, and suggesting those methods and practices that can minimise the anthropogenic impacts on the environment, from reducing energy and water consumption to mitigating the risks arising from land interventions. In other words, they could freely implement the value of serving the community that is inherent in geoscientific activity. Unfortunately, in a logic of apparently irreconcilable opposition between public and private good, the sense of the individual and their social responsibility lose their value and an irremediable inner fracture between ethical conduct and corporate interest is produced.

But, it is also true that companies cannot lose social credibility and that the adoption of environmentally friendly and socio-sustainable policies could certainly improve their image in local communities, which would welcome the proposed solutions more favourably. For this reason, geoscientists working in these fields can become promoters of ethical and innovative solutions, which can also prove to be of great benefit to the company they work for, as they can enhance its social reputation. In this case, the common good will coincide with the advantage of the company.

Another important aspect that presupposes the social responsibility of the geoscientist is communication of knowledge to the different components of society. In all cases where there are no restrictions to guarantee intellectual and/or industrial property, it is important that research results are shared both with expert users directly interested in their application and with non-expert users, but in a way that takes into account the level of knowledge, interests, needs and concerns of the various stakeholders. It is also important that they are as accessible, user-friendly, contextualised and accompanied by explanatory information tailored to the different end users as possible. In fact, openly distributing data (especially so-called 'raw' data) or the results of scientific processing without concern for the final recipient of that information may create the conditions for misunderstandings, misinterpretations or the distorted use of the information itself. For example, in the field of geological hazards data relating to an area cannot be transferred tout court to the public without due clarification, as an incorrect reading of them could give rise to false alarmism or excessive reassurance. Similar care must also be taken when communicating scientific knowledge to political decision-makers or public bodies, which will have to use that information for the operational management of the territory. It is only by taking such precautions that one can foster the development of constructive and responsible interactions between the scientific communities and society, which are instrumental in establishing relationships of trust and respect in the pursuit of common goals (Stewart et al., 2017).

The growing awareness of their social role is also prompting geoscientists to raise alarms that affect the entire planet. In addition to giving a picture of the worrying state of evolution of some environmental parameters, they are also able to suggest some possible solutions to modify evolutionary trends, to limit the impacts of anthropogenic global changes (Jouffray et al., 2020; Ripple et al., 2017, 2020, 2021). With increasing accuracy, studies show a reduction in freshwater availability per capita on Earth, an increase in the number of anoxic coastal areas due to the release of fertilisers, other chemical nutrients, and fossil fuels into the water, an increase in deforested zones especially in tropical areas, a significant reduction in biodiversity, an alteration of biogeochemical cycles,[1] an increase in the concentration of carbon dioxide (CO_2), methane (CH_4) and nitrous oxide (N_2O) in the atmosphere,

---

[1] The biogeochemical cycles (of carbon, oxygen, phosphorus, sulphur, nitrogen, water) are characterised by dynamically balanced processes consisting of essentially circular pathways, through

and an increase in acidification of ocean waters, global average surface temperature, world population, energy consumption, and air transport. The recommendations and concrete actions proposed by geoscientists in the energy and economic sectors to reduce greenhouse gas emissions, protect ecosystems or reduce land consumption are essential to achieve not only the goals of ecological sustainability, but also those of social and economic justice, which are closely linked to the use of resources and land protection, and which governments must include among their measures, adjusting their political priorities.

Geoscientists are also able to define the key issues and determine the most important indicators to quantify the effectiveness of interventions. Moreover, they have the tools to act within the scientific community itself, motivating and creating awareness of actions to be taken, being able to develop effective methods and techniques to improve sustainability, adaptive capacity, risk defence and resilience of society (Gill & Bullough, 2017).

But, their actions to raise awareness of the protection and enhancement of the geo-biological, ecological, and landscape heritage go even further, taking on the value of preserving the sense of place, the historical-environmental identity, and the economic and cultural vocation of the territory, which should also be understood as a place of functional interaction between human communities and nature. Knowing the territory, its natural beauty, its geological fragility, the history of the use of its natural resources and the ways in which humans interact with local ecosystems does not only have a didactic, informative or touristic value, but it means providing society with a key to understanding the balance with nature that the inhabitants of a place have painstakingly built over time and that still structure their cultural identity.

Moreover, geoscientists can play a key role in bringing the population closer to scientific knowledge by involving them in educational activities or directly in scientific research through *citizen science* (Eitzel et al., 2017), activities in which citizens voluntarily participate as non-professional scientists in data collection and analysis, development of technologies or studies of natural phenomena (Bonney et al., 2014; De Rubeis et al., 2015; Riesch & Potter, 2014).

Citizen science is based on the idea that knowledge and scientific communication are not a one-way street proceeding from the scientist to the population but activities of mutual exchange, of co-creation of knowledge, since citizens can also provide scientists with valuable support (Paul et al., 2018; Vayena & Tasioulas, 2015). In the field of natural risks, the information that citizens can provide, for example on how they felt a seismic event, is often useful in complementing the studies of seismologists (De Rubeis et al., 2015), and this active involvement of citizens in the scientific enterprise has a great educational and ethical meaning, as it facilitates the understanding of phenomena, generating awareness, and, consequently, responsibility. It is crucial that geoscientists are aware that their work can contribute to building a more informed society and also being more actively involved in political issues, since a

---

which the chemical elements essential for biological life are continuously transferred from the atmosphere, hydrosphere and lithosphere to the organisms in the ecosystems and vice versa.

greater knowledge of phenomena increases the citizens' ability to evaluate the political choices adopted in the management of the territory and increases awareness in their possibilities of action.

If, therefore, geoscientists are responsible for the 'transfer' of knowledge and the formation of a scientific culture in society, citizens, for their part, have the difficult task of getting information from reliable and certified sources. The proliferation of pseudo-scientific or false information on geo-environmental issues, propagandised as plausible scientific hypotheses, produces double damage to the community; on one hand, it creates confusion and feeds distrust in the possibilities of science as a support to daily life, and on the other hand, it inevitably increases social vulnerability to any change or risk.

As stated above, the citizens' right to knowledge and correct information has, as its ethical implication, the duty to seek this knowledge from those who produce it, following strict control and verification mechanisms (such as universities, research institutes, centres of excellence) or from those who are responsible for translating it into actions on the territory (such as government bodies). From this point of view, politicians and local authorities, as well as those working in the communication and information sectors, also have a duty to avoid the dissemination of news and data within society that have not been collected and analysed according to established scientific methods, and to avoid inviting into public debates those who do not have qualifications and expertise in scientific issues. In the debate on anthropogenic climate change, it is now common to see media reports that equate the denialist views of a very small number of scientists, often with no specific expertise on climate issues, with the scientific findings on which almost all of the international scientific community now agrees (Lynas et al., 2021; Oreskes, 2014; Petersen et al., 2019).

In a globalised world, every human agent has clear responsibilities in the circumscribed area in which they live, but they cannot think that their sphere of influence is limited to that. Geoethics calls upon a responsibility of all towards all. The effects of an individual's personal behaviour add up to those produced by other individuals, with important consequences for the complex dynamics of the system. If we think of any object used in our daily lives, we cannot fail to consider that it is the product of the industrial transformation of natural resources. For example, the smartphone is the product of an industrial transformation and assembly chain based on metals and rare earths. The extraction of rare earths is often carried out by exploiting child labour or trading for armaments which fuel local conflicts in unfortunate areas of the world. Knowing the background of industrial production can certainly increase consumer awareness and, in time, foster a change in individual consumption behaviour that stops feeding a market and labour system based on unacceptable inequalities and inequities.

The so-called 'NIMBY syndrome' (an acronym for the expression 'Not In My Back Yard'), which characterises those citizens who are against the construction of industrial plants and infrastructures impactful on the territory where they live (such as landfills, geothermal plants or oil and gas extraction plants, oil pipelines, railway lines, motorways) is also an example of how the lack of knowledge of an environmental problem creates prejudicial hostility in citizens. Citizens must be able to understand that it is possible to build plants and facilities according to the best construction and safety and health defense standards, provided that they are subjected to rigorous public controls. But, instead of insisting on efficient controls, it happens that this hostility simply takes the form of a request to build these facilities elsewhere. And even if everyone understands the public usefulness of such facilities, if only because they are functional to today's organisational system of society and our current lifestyles, no one is willing to take on part of the collective burden. The Nimby syndrome even becomes opposition to the implementation of works or activities within one's own country and the transfer of economic and environmental problems to the population of countries that have less weight and bargaining power to demand control, safety, and health from their governments. Even the use of fossil fuels, which still represents the preponderant part of the energy mix in western societies, has similar implications. In some countries, public opinion does not want oil and gas to be extracted on their territory, even in the presence of advanced national legislation, modern technologies, and an efficient system of controls by public bodies, preferring other countries take charge of this task, even in the absence of the necessary safety and technological reliability requirements, or in the total absence of adequate legal and judicial frameworks to guarantee the rights of the population and effective environmental protection.

Therefore, any human agent aware of their responsibilities has first and foremost the duty to place their choices and actions in a broader and more complex context than their own narrow experience. Their attitudes of prejudicial hostility do nothing but irresponsibly shift the burden of meeting energy needs, providing mineral resources or disposing of waste onto weaker individuals.

Every individual has a geoethical duty to develop an adequate level of awareness of global issues, a feeling of solidarity and commonality towards other human beings, and intransigence towards all forms of iniquity and human exploitation and destruction of the territory. But, the sharing of responsibility between different individuals, groups, and communities cannot be separated from the identification of a common system of reference values, functional to a more effective cooperation in the management of complex problems.

4.4.4 Environmental Responsibility

Nature does not need human beings to continue to exist and evolve (Pievani, 2019), but human beings do need to establish a relationship with nature that is capable of guaranteeing them health, safety and harmony.

It is now undeniable that anthropogenic impacts on the abiotic and biotic forms and processes of our planet are capable of often irreversibly modifying ecological systems, to the point that some scientists believe that the Earth system is now an inextricable evolutionary tangle of human and non-human realities (Ellis & Haff, 2009), so permeated that they now appear only in the form of humanised nature, anthropogenic biomes,[2] modified by human action (Ellis & Ramankutty, 2008), sometimes called 'anthromes'.

The responsibility of geoscientists towards the planet stems from their knowledge of Earth's forms and dynamics, including hazards accompanying certain phenomena, the limits of the albeit transitory balances governing ecosystems, the beauty and functionality of geodiversity and biodiversity, and the richness of natural resources. Their knowledge, experience, and professional skills are essential to protect natural environments, manage resources wisely, minimise negative impacts on the environment and transfer the scientific, educational, cultural and aesthetic value of natural diversity to future generations. Society should rely on geoscientists to identify new ways of social and economic development that take into account the functioning and vulnerabilities of ecosystems and establish a functional relationship with Earth that ensures quality and respect for life (human, animal and plant) and non-living elements (water, air, soil, rocks).

By emphasising the individual and collective responsibilities of the human agent, geoethics dismantles the passive or opportunistic acceptance of anthropogenic change by those who, due to short-sightedness, ignorance or selfishness, do not want to give up the satisfaction of their desires and needs, even at the cost of profound environmental impacts, often considered inevitable collateral damage rather than recognised for the actual nature of their actions, based on economic and cultural paradigms that are now obsolete, if not downright dangerous.

In the third millennium, one cannot consume the planet's resources without, at the same time, taking care not to exceed the ecological thresholds capable of ensuring human survival on Earth (Rockström et al., 2009; Steffen et al., 2015). This implies proactive action to change everyday lifestyles, political activism that affects the issues, and measures developed by legislators and decision-makers due to a changing ecological awareness. In modern societies there is a need to assign responsibility equally to individuals, to share it within an ideal framework that is also a reference for individual gestures, organised according to a succession of actions aimed at achieving a broader and more articulated ecological awareness, spatially and temporally.

An apparently innocuous individual behaviour, such as excessive meat consumption, not only increases a person's vulnerability to specific diseases but also represents the discretisation of a broader collective action which, within our economic system, causes heavy damages to the environment, the depletion of drinking water reservoirs, increased emissions of atmospheric greenhouse gases and the loss of forest areas. Similarly, the even modest spill of toxic compounds may seem an action limited to the area in which it occurs, the effect of the low consciousness of an individual.

---

[2] 'Biomes are complex environmental systems of wide geographical extent, consisting of a set of ecosystems, whose animal and plant communities have achieved, in a given area of the Earth's surface, a relative stability in relation to environmental conditions. Each biome is characterised primarily by the climatic conditions of the region and by particular vegetation that hosts a typical fauna (set of animal species)' (translated in English from: https://www.sapere.it/sapere/strumenti/studiafacile/biologia/Organismi-e-ambiente/La-biosfera-e-i-biomi/I-biomi.html). See also: https://www.sciencedirect.com/topics/earth-and-planetary-sciences/biome. Accessed 29 March 2022.

Unfortunately, the soil and water pollution caused by such a spill can extend well beyond the local area, contaminating groundwater, surrounding rivers, and eventually the sea. The most frequent consequences are eutrophication and oxygen depletion of waters, the death of fish fauna, with major repercussions on the local economy. In the most serious cases, the territory can be poisoned to such an extent as to produce toxic or lethal effects on the health of the population involved.

Our potential or actual interaction with the natural system can be very invasive, to the extent that decisions made in a territory can have repercussions even kilometres or years away from the place or moment they were implemented.

Certainly, in relation to environmental degradation or resource exploitation, there are great disparities in terms of current and historical responsibilities within human societies and between the individuals who compose them (Bonneuil & Fressoz, 2013). Generally speaking, richer societies are also responsible for the largest consumption of mineral, energy, water and animal resources on the planet, while poorer societies often follow lifestyles that are more sustainable and overall less impactful on natural environments.

For this reason, states are called upon to distribute their responsibilities towards the planet fairly and in solidarity, within and among themselves, according to the severity of the negative consequences that their decisions have had and have on natural systems. Failure to diversify the responsibility of each state according to the causal weight it has produced in global environmental terms can also pose a major threat to the social and political stability of present and future societies, as a harbinger of further inequalities, disputes, and ecological disasters.

References

Andrews, G. C. (2014). *Canadian professional engineering and geoscience practice and ethics* (5th ed., p. 275). Nelson College Indigenous, ISBN 978–0176509903.

Abbott, D. M. (2017). Some fundamental issues in geoethics. *Annals of Geophysics, 60*, Fast Track 7: Geoethics at the heart of all geoscience. doi:https://doi.org/10.4401/ag-7407

Bohle, M., & Di Capua, G. (2019). Setting the scene. In M. Bohle (Ed.), *Exploring geoethics—Ethical implications, societal contexts, and professional obligations of the geosciences* (pp. 1–24). Palgrave Pivot, Cham. https://doi.org/10.1007/978-3-030-12010-8_1

Bonneuil, C., & Fressoz, J. -B. (2013). *L'Evénement Anthropocène - La Terre, l'histoire et nous* (p. 320). Seuil, Paris, ISBN 978–2021135008.

Bonney, R., Shirk, J. L., Phillips, T. B., Wiggins, A., Ballard, H. L., Miller-Rushing, A. J., & Parrish, J. K. (2014). Next steps for citizen science. *Science, 343*(6178), 1436–1437. https://doi.org/10.1126/science.1251554

De Rubeis, V., Sbarra, P., Sebaste, B., & Tosi, P. (2015). Earthquake ethics through scientific knowledge, historical memory and societal awareness: the experience of direct Internet information. In S. Peppoloni & G. Di Capua (Eds), Geoethics—The role and responsibility of geoscientists. *Geological Society of London, Special Publications, 419* 103–110. https://doi.org/10.1144/SP419.7

Doglioni, C., & Peppoloni, S. (2016). *Pianeta Terra: una storia non finita* (p. 160). Il Mulino, Bologna, ISBN 978–8815263766.

Eastin, J. (2018). Climate change and gender equality in developing states. *World Development, 107*, 289–305. https://doi.org/10.1016/j.worlddev.2018.02.021

Eitzel, M. V; Cappadonna, J. L; Santos-Lang, C; Duerr, R. E; Virapongse, A; West, S. E; Kyba, C. C. M., Bowser, A., Cooper, C. B., Sforzi, A., Metcalfe, A. N., Harris, E. S., Thiel, M., Haklay, M., Ponciano, L., Roche, J., Ceccaroni, L., Shilling, F. M., Dörler, D., Heigl, F., Kiessling, T., Davis, B. Y., & Jiang, Q. (2017). Citizen science terminology matters: Exploring key terms. *Citizen Science: Theory and Practice, 2*(1), 1. https://doi.org/10.5334/cstp.96

Ellis, E. C., & Haff, P. K. (2009). Earth science in the anthropocene: New epoch, new paradigm, new responsibilities. *EOS Transactions, 90*(49), 473. https://doi.org/10.1029/2009EO490006

Ellis, E. C., & Ramankutty, N. (2008). Putting people in the map: Anthropogenic biomes of the world. *Frontiers in Ecology and the Environment, 6*(8), 439–447. https://doi.org/10.1890/070062

Franco, V. (2015). *Reponsabilità—Figure e metamorfosi di un concetto* (p. XII+196). Donzelli editore, Roma, ISBN 978–8868431938.

Gill, J. C., & Bullough, F. (2017). Geoscience engagement in global development frameworks. *Annals of Geophysics, 60*, Fast Track 7: Geoethics at the heart of all geoscience. doi:https://doi.org/10.4401/ag-7460

Jonas, H. (1979). Das Prinzip Verantwortung: Versuch einer Ethik für die technologische Zivilisation. Suhrkamp, Frankfurt/M. *The imperative of responsibility: In search of ethics for the technological age* (translation of Das Prinzip Verantwortung) trans. Hans Jonas and David Herr (1979). ISBN 0–226–40597–4 (University of Chicago Press, 1984), ISBN 0–226–40596–6.

Jouffray, J. B., Blasiak, R., Norström, A. V., Österblom, H., & Nyström, M. (2020). The Blue Acceleration: The trajectory of human expansion into the ocean. *Perspective, 2*(1), 43–54. https://doi.org/10.1016/j.oneear.2019.12.016

Lynas, M., Houlton, B. Z., & Perry, S. (2021). Greater than 99% consensus on human caused climate change in the peer-reviewed scientific literature. *Environmental Research Letters, 16*. https://doi.org/10.1088/1748-9326/ac2966

Mayer, T. (2015). Research integrity: The bedrock of the geosciences. In M. Wyss & S. Peppoloni (Eds.), *Geoethics: Ethical challenges and case studies in earth sciences* (pp. 71–81). Elsevier. https://doi.org/10.1016/B978-0-12-799935-7.00007-1

Mogk, D. (2017). Geoethics and professionalism: The responsible conduct of scientists. *Annals of Geophysics, 60*, Fast Track 7: Geoethics at the heart of all geoscience. doi:https://doi.org/10.4401/ag-7584

Mogk, D., Geissman, J. W., & Bruckner, M. Z. (2017). Teaching geoethics across the geoscience curriculum. Why, when, what, how, and where? In L. C. Gundersen (Ed.), *Scientific Integrity and Ethics: With Applications to the GeosciencesSpecial Publications 73*, 231–265. American Geophysical Union. https://doi.org/10.1002/9781119067825.ch13

Oreskes, N. (2014). The scientific consensus on climate change. *Science, 306*(5702), 1686. https://doi.org/10.1126/science.1103618

Ballesteros-Cánovas, J. A., Bhusal, J., Cieslik, K., Clark, J., Dugar, S., Hannah, D. M., Stoffel, M., Dewulf, A., Dhital, M. R., Liu, W., Nayaval, J., Neupane, B., Schiller, A., Smith, P. J., & Supper, R. (2018). Citizen science for hydrological risk reduction and resilience building. *Wiley Interdisciplinary Reviews: Water, 5*(1). https://doi.org/10.1002/wat2.1262

Peppoloni, S., Bilham, N., & Di Capua, G. (2019). Contemporary geoethics within the geosciences. In M. Bohle (Ed.), *Exploring geoethics—Ethical implications, societal contexts, and professional obligations of the geosciences* (pp. 25–70), Palgrave Pivot. https://doi.org/10.1007/978-3-030-12010-8_2

Peppoloni, S., Bobrowsky, P., & Di Capua, G. (2015). Geoethics: A challenge for research integrity in geosciences. In N. Steneck, M. Anderson, S. Kleinert, & T. Mayer (Eds.), *Integrity in the global research Arena, World scientific* (pp. 287–294). https://doi.org/10.1142/9789814632393_0035

Peppoloni, S., & Di Capua, G. (2015). The meaning of geoethics. In M. Wyss & S. Peppoloni (Eds.), *Geoethics: Ethical challenges and case studies in earth sciences* (pp. 3–14). Elsevier . https://doi.org/10.1016/B978-0-12-799935-7.00001-0

Peppoloni, S., & Di Capua, G. (2016). Geoethics: Ethical, social, and cultural values in geosciences research, practice, and education. In G. R. Wessel, & J. K. Greenberg (Eds.), Geoscience for the public good and global development: Toward a sustainable future. *Geological Society of America, Special Paper, 520*, 17–21. https://doi.org/10.1130/2016.2520(03

Peppoloni, S., & Di Capua, G. (2017). Geoethics: Ethical, social and cultural implications in geosciences. *Annals of Geophysics, 60*, Fast Track 7: Geoethics at the heart of all geoscience. https://doi.org/10.4401/ag-7473

Peppoloni, S., & Di Capua, G. (2018). Ethics. In P. T. Bobrowsky & B. Marker (Eds), *Encyclopedia of Engineering Geology*, Encyclopedia of Earth Sciences Series, Springer. https://doi.org/10.1007/978-3-319-12127-7_115-1

Peppoloni, S., & Di Capua, G. (2020). Geoethics as global ethics to face grand challenges for humanity. In G. Di Capua, P. T. Bobrowsky, S. W. Kieffer & C. Palinkas (Eds.), *Geoethics: Status and Future Perspectives, Geological Society of London, Special Publications, 508*, 13–29. https://doi.org/10.1144/SP508-2020-146

Peppoloni, S., & Di Capua, G. (2021). Geoethics to start up a pedagogical and political path towards future sustainable societies. *Sustainability, 13*(18), 10024. https://doi.org/10.3390/su131810024

Petersen, A. M., Vincent, E. M., & Westerling, A. L. (2019). Discrepancy in scientific authority and media visibility of climate change scientists and contrarians. *Nature Communications, 10*, 3502. https://doi.org/10.1038/s41467-019-09959-4

Pievani T. (2019). *La Terra dopo di noi* (p. 184). Contrasto, Roma, ISBN 978–8869657887.

Riesch, H., & Potter, C. (2014). Citizen science as seen by scientists: Methodological, epistemological and ethical dimensions. *Public Understanding of Science, 23*(1), 107–120. https://doi.org/10.1177/0963662513497324

Ripple, W. J., Wolf, C., Newsome, T. M., Galetti, M., Alamgir, M., Crist, E., Mahmoud, M. I., & Laurance, W. F. (2017). World scientists' warning to humanity: A second notice. *BioScience, 67*(12), 1026–1028. https://doi.org/10.1093/biosci/bix125

Ripple, W. J., Wolf, C., Newsome, T. M., Barnard, P., & Moomaw, W. R. (2020). World scientists' warning of a climate emergency. *BioScience, 70*(1), 8–12. https://doi.org/10.1093/biosci/biz088

Ripple, W. J., Wolf, C., Newsome, T. M., Gregg, J. W., Lenton, T. M., Palomo, I., Eikelboom, J., Law, B. E., Huq, S., Duffy, P. B., & Rockström, J. (2021). World scientists' warning of a climate emergency 2021. *BioScience, 71*(9), 894–898. https://doi.org/10.1093/biosci/biab079

Rockström, J., Steffen, W., Noone, K., Persson, A., Chapin, F. S., Lambin, E. F., Lenton, T. M., Scheffer, M., Folke, C., Schellnhuber, H. J., Nykvist, B., de Wit, C. A., Hughes, T., van der Leeuw, S., Rodhe, H., Sörlin, S., Snyder, P. K., Costanza, R., Svedin, U., Falkenmark, M., Karlberg, L., Corell, R. W., Fabry, V. J., Hansen, J., Walker, B., Liverman, D., Richardson, R., Crutzen, P., & Foley, J. A. (2009). A safe operating space for humanity. *Nature, 461*(7263), 472–475. https://doi.org/10.1038/461472a

Steffen, W., Richardson, K., Rockström, J., Cornellingo, S. E., Bennettreinette, Fetzer, I., Bennett, E. M., Biggs, R., Carpenter, S. R., de Vries, W., de Wit, C. A., Folke, C., Gerten, D., Heinke, J., Mace, G. M.,Persson, L. M., Ramanathan, V., Reyers, B., & Sörlin, S. (2015). Planetary boundaries: Guiding human development on a changing planet. *Science, 347*(6223), 1259855–1259855. https://doi.org/10.1126/science.1259855

Stewart, I. S., Ickert, J., & Lacassin, R. (2017). Communication seismic risk: The geoethical challenges of a people-centred, participatory approach. *Annals of Geophysics, 60*, Fast Track 7: Geoethics at the heart of all geoscience. https://doi.org/10.4401/ag-7593

Vayena, E., & Tasioulas, J. (2015). "We the Scientists": A human right to citizen science. *Philosophy & Technology, 28*(3), 479–485. https://doi.org/10.1007/s13347-015-0204-0

Chapter 5
The Advantage of Geoethical Action

5.1 Why Should We Behave Ethically?

The answer to why one should behave ethically may seem obvious: 'because otherwise one will be punished', or 'because I was brought up to behave well', or even 'because one would live in a better way' or 'because it is right, as it could not be so'. But, in its simplicity, this question can cause disorientation and raise doubts. Personal motivations and resulting justifications are rooted in the whole of moral references that each person builds up over the course of their life, on the basis of elements such as family and school upbringing, the attending of specific groups and communities, the social climate breathed in, and the way in which the progressive re-elaboration of one's experiences has taken place. These references are mobile, changing over time, as a function of changes in our individual perception of what is right on the basis of changing boundary conditions. Therefore, moral references are the response to a continuous process of interaction/verification of one's ideas and actions with one's inner dimension and the external world. Over time, the inputs acting on the individual faced with decisions to be made change, as do their existential circumstances. The problems to be addressed also change in relation to cultural, technological, and economic changes in society, and individual and social sensitivities may lead to a different perception of issues such as civil rights, the environment, and justice. The globalised and technologised world also confronts individuals with ever new questions and dilemmas in their interconnections.

It is even more confusing to ask why we should behave ethically when others are unwilling to do so, or even question our own values by their actions, or go against the guidelines deemed right or appropriate by the social group to which we belong, or, finally, cause harm to others by their behaviours. In such cases, the functional balance of social relations and the respect between human beings also breaks down. Ultimately, the reciprocity on which the unwritten contract that governs the interactions between people and groups is based falls apart.

Let us ask ourselves whether we are willing to behave ethically even if there is no reciprocity and, thus, no recognition of value on the part of all moral subjects. In fact, we are asking ourselves what is the acceptable limit for ethical behaviours, beyond which the reference moral frame no longer supports our choices. From this point of view, ethics does not have a monolithic structure but finds variable application on the basis of contexts and the changing hierarchical scale of values adopted, without prejudice to the reference principles and values that underlie its conceptual framework.

Ethical behaviour may mean something different to an individual than to a group of people or an entire society. After all, the diversity of human sensitivities and cultures is such that while to standardise individual visions and behaviours represents a serious impoverishment of the variety and richness of the points of view and qualities of humanity itself, it is a source of conflict, inequality, and difficulties in cooperation. In any case, there remains an objective need to find that reasonable balance of values, on the basis of which to give homogeneous form to behaviours capable of renewing interaction between human beings, each characterised by personal and legitimate interests, and between the human being and the Earth system. The current historical moment seems very favourable for bringing individual and social demands closer together, for aligning the motivations of personal ethical choices with those of the whole of society, in a changed framework of perception of the self in relation to the social-ecological system in which it is immersed.

From this perspective, behaving ethically means implementing one's own moral dimension, while at the same time contributing to the implementation of something greater. Thus, the ethical behaviour of the individual is the quantum of a social action in which the individual plays an active and qualified role, contributing to changing the world and achieving a goal or common good. In this case, behaving ethically will also determine a better functioning of the system, understood not only as the complex of social interactions in which one is inserted but also the dimension extended to include the entire planet, nothing more admirable, motivating, and useful that can happen in the life of an individual.

This condition implies that we perceive ourselves as earthlings, inhabitants of Earth, that is to say of that complex physical-chemical-biological system but also social and technological, in perpetual disequilibrium and in continuous search of transitory equilibriums, of which each human being is an integral, qualifying, and acting part. A humanity that tends towards this widespread awareness can pragmatically answer the question initially posed with, 'it is right, also because it is convenient for everyone, first and foremost for humanity'. Behaving ethically allows each human being to experience the Earth system in a way that is more functional to their security and health (and therefore to their happiness), guaranteeing dignified living conditions also for future generations, not necessarily the closest ones, so that they in turn can freely and responsibly determine what their present will be.

5.2 Towards a New Political Agenda

As early as in the second half of the eighteenth and nineteenth centuries, warnings about the anthropogenic deterioration of natural environments and the associated risks emerged as a response to economic development and the growing industrialisation of some areas of the planet. As Bonneuil and Fressoz (2013) point out, at that time scientists and intellectuals were already talking about the 'power of man' in modifying the 'face of the Earth' (Buffon, 1707–1788), the exploitation of the planet (Saint-Simon, 1760–1825), the future destruction of forests and huge carbon dioxide emissions (Eugène Huzar, 1820–1890), the human impact on biogeochemical cycles (Vladimir Vernadsky, 1863–1945). Lewis and Maslin (2018) highlight that even in the seventeenth century John Evelyn (1620–1706) wrote about London in apocalyptic terms, referring to the bad air quality, suggesting to the King of England the planting of trees as a measure to counter pollution. Similarly, in 1827 Jean Baptiste Joseph Fourier (1768–1830) spoke of the 'material deterioration of the planet' and in 1876 Friedrich Engels (1820–1895) spoke of man-made environmental disasters as an effect of the capitalist economy. In the twentieth century, scientists begin to delineate and quantify, with increasing precision, the impact of anthropogenic actions on the Earth system (Jouffray et al., 2020; Ripple et al., 2020, 2021). Environmentalist movements begin to develop, and society and the political world become increasingly aware that local and global changes in natural environments are the dark side of a global economic system (Bonneuil & Fressoz, 2013) that does not take into account planetary boundaries.

This progressive awareness shows that the current tension towards an ethical dimension to the environment began as a historical necessity, the result of a slow process over the last three hundred years. In more recent times, this process has undoubtedly accelerated as a result of increased concern about uncertain social and natural conditions that may arise in the near future.

In this ideal path, the boosts of the international youth movement 'Fridays for Future'[1] are also well framed. The political demands of this movement take the form of a call for a complete energy transition to clean and renewable energies, an immediate halt to exploration for fossil fuels and the complete elimination of subsidies for their use, concrete help for the victims of climate change, the causes of which are mainly traced to the economic models of the richest countries and whose effects are suffered mainly by the poorest populations in the world.

Although these demands clash with the global economic and political framework, there is no doubt that they demonstrate the global moral development underway, which places at the centre of the world public debate the imperative of a change, sometimes radical, in lifestyles, economic paradigms, cultural structures and social organisations. It is on this terrain that these new generations want to measure the ethical profile of the current ruling classes in order to decide who should govern human societies in the future.

---

[1] https://www.fridaysforfuture.org/. Accessed 29 March 2022.

The difference with the youth movements of the past seems to be the determination to put environmental issues right at the centre of the political agendas, reconsidering all the other fundamental issues for society as a function of a better relationship between human beings and the planet. Anthropogenic climate change (but we could also speak more broadly of an environmental issue) is not simply one of the topics of political debate but is considered the central issue on which the questions of social inequalities, welfare, justice, territorial government, management of natural resources, and international conflicts are grafted. Addressing the so-called 'climate emergency' means becoming aware of it, redefining society's hierarchy of ethical reference values within a changed perception of reality, and finally giving substance to one's vision of existence in order to resolve the new problems that have emerged.

5.3   Moral Development and Ecological Action

The ecological concerns of some forward-looking eighteenth- and nineteenth-century intellectuals have now materialised in a widespread perception of danger in world public opinion. The action of youth movements is proving capable of creating networks of cooperation and opening channels of communication with authoritative scientists and influential politicians. This is to ensure that environmental issues are urgently included in the action programmes of governments, according to a vision that is motivated not only by fear but also by an increasingly strong idealistic drive to change attitudes and ways of living on the planet. This cannot but be considered a form of human moral progress.

In this respect, Lawrence Kohlberg (1927–1987) also provides an interesting reading key to geoethics (Bohle & Marone, 2019; Marone & Peppoloni, 2017). He elaborates a theory of moral development as growth of personal awareness, placing at the basis of his reflections the human beings who learn, are motivated, grow culturally, learn to move in the universally existing social institutions in which they participate (such as the family, the professional group, society) and also by imitating behavioural models of reference, through which the personal moral maturity degree is tested. All this can shape the individual's ethical dimension over time, which develops across cultures. Kohlberg defines three levels of moral development, each of which is subdivided into two stages, making a total of six stages to be understood as levels of moral adequacy and used to explain the development of ethical reasoning (Kohlberg, 1981).

In the first (pre-conventional) level, where the morality of actions is judged on the basis of their direct consequences, there is a first (heteronomous) stage. This stage is characterised by an individual who makes choices on the basis of external orders, hence ignoring motivation and ends, and the verification of the rightness of the choice is in relation to a punishment mechanism, i.e., the individual is guided by blind egoism that stems from obedience out of fear of punishment. In the next stage (stage of individualism, of instrumental purpose, of exchange), the individual pursues their own motivations and goals out of instrumental egoism, letting others do the same.

In the second (conventional) level, the morality of actions is judged on the basis of society's expectations and visions, and moral adherence to the established rules occurs more out of observance than out of conviction. Within this level there is the third stage, where the interpersonal relationships of individuals within groups come into play, and moral development is shaped on the basis of these social relationships. Values such as mutual respect, loyalty, trust, cooperation and altruism become important. Still at this level, the fourth stage considers morality, referring it to the broader sphere of society, whereby moral development is essentially assessed on the basis of the individual's compliance with the law, which guarantees the maintenance of social order and legal authority.

At the third (post-conventional) level, the morality of the individual may take precedence over the morality of society, in the knowledge that the latter may vary and may be established, for example, in the form of agreements between individuals. In fact, in the fifth stage the perimeter of morality is defined within a social contract that establishes those individual values and rights considered to be the basis of the law. Adherence to the social system in which one participates occurs by free choice. In this stage, one becomes aware of the foundations of the law whereas, in the previous stage, one was only concerned with behaving in accordance with it. Finally, it is only in the sixth stage that one comes to develop an enriched conscience of ethical principles that the individual considers universally applicable and that guides all their choices, even in cases where conflicts arise with the law that regulate society. Therefore, these principles go beyond the law. At this stage, action is never a means to achieve a goal, and it does not serve to avoid punishment or to comply with social laws, but it is right in itself, conforming to principles that the individual experiences as constituent parts of one's own being.

It is at this last stage of moral development that principles such as respect for life, freedom, justice, equality, fairness and other fundamental human rights become embedded in the intentions and actions of each individual, who has reached the highest level of ethics and responsibility.

From the point of view of geoethics, environmental protection and respect for geodiversity and biodiversity are the foundations of a new ecological sensitivity, which stems from a historical process that has also been a path of moral development according to the stages elaborated by Kohlberg. This process 'imposes' on individual life choices that, although not prescribed by laws or not fully understood in the complexity of their implications, are considered intrinsically just, because they are inherent to an innate feeling, to that ontological foundation that makes us nature. This gives an extraordinary ethical force to action, enlightening even the most resistant political pathways to the epochal changes demanded by the younger generations. In this case, therefore, ethical behaviour stems from the need to adhere to one's own identity as nature, it consists in a non-violent reaction to political sloth and inertia, and it takes place in caring for the common home.

Geoethical thought has the merit of providing the motivations for ecological action from historical, philosophical, and scientific perspectives, outlining a common moral matrix between the individual and social spheres, placing the individual at the centre of a systemic and multidisciplinary response to anthropogenic problems that affect the entire Earth system as an effect of a general human ethical development.

References

Bohle, M., & Marone, E. (2019). Humanistic geosciences and the planetary human niche. In M. Bohle (Ed.), *Exploring geoethics—Ethical implications, societal contexts, and professional obligations of the geosciences* (pp. 137–164). Palgrave Pivot. https://doi.org/10.1007/978-3-030-120 10-8_4

Bonneuil, C., & Fressoz, J.-B. (2013). *L'Evénement Anthropocène - La Terre, l'histoire et nous* (p. 320). Seuil. ISBN 978-2021135008.

Jouffray, J.-B., Blasiak, R., Norström, A. V., Österblom, H., & Nyström, M. (2020). The blue acceleration: The trajectory of human expansion into the ocean. *Perspective, 2*(1), 43–54. https://doi.org/10.1016/j.oneear.2019.12.016

Kohlberg, L. (1981). *The philosophy of moral development: Moral stages and the idea of justice* (Essays on Moral Development, Vol. 1, p. 441). Harper & Row. ISBN 978-0060647605.

Lewis, S. L., & Maslin, M. A. (2018). *The human planet: How we created the Anthropocene* (p. 480). Pelican. ISBN 978-0241280881.

Marone, E., & Peppoloni, S. (2017). Ethical dilemmas in geosciences. We can ask, but, can we answer? *Annals of Geophysics, 60, Fast Track 7: Geoethics at the heart of all geoscience.* https://doi.org/10.4401/ag-7445

Ripple, W. J., Wolf, C., Newsome, T. M., Barnard , P., & Moomaw, W. R. (2020). World scientists' warning of a climate emergency. *BioScience, 70*(1), 8–12. https://doi.org/10.1093/biosci/biz088

Ripple, W. J., Wolf, C., Newsome, T. M., Gregg, J. W., Lenton, T. M., Palomo, I., Eikelboom, J. A. J., Law, B. A., Huq, S., Duffy, P. B., & Rockström, J. (2021). World scientists' warning of a climate emergency 2021. *BioScience, 71*(9), 894–898. https://doi.org/10.1093/biosci/biab079

Chapter 6
Ethical Problems and Dilemmas in the Geosciences

6.1 The Search for a Functional Balance

Any decision may be made impulsively or may mature after weighing up a range of possibilities. Deciding what is the right or acceptable thing to do in a situation open to multiple solutions implies in the simplest cases choosing between two or more alternatives, one of which is the best. The choice, besides depending on the operational context, will be based on intuitions, reflections, experiences, knowledge, emotions, interests, expectations, priorities, calculations, individual, social, cultural, economic, and scientific values of reference for each person (Elster, 2009; Simon, 1985). The decision-making process is structured in a sequence of mental operations that the individual agent carries out in order to choose between different alternatives, having to define a single objective of his actions (Lindsay & Norman, 1972).

Complete and accurate knowledge of the situation and the problem to be addressed, careful analysis of the possible consequences and the appropriate competence in identifying solutions are indispensable foundations for making informed and responsible decisions.

If one solution is clearly better than the others, then the decision to be made is relatively simple, since the advantages of that choice over the others are clear. But, as often happens in the practice of geosciences, one is faced with dilemmas (Bilham, 2015) whereby several solutions exist, but each of them also brings with it negative consequences for the human community or the environment involved (Marone & Peppoloni, 2017). In these cases a 'universally right' choice is not possible and one finds oneself in the difficult position of deciding among them.

An ethical dilemma does not have an ideal, universally acceptable solution but offers a solution that can be considered the best possible for that specific economic, social, cultural and environmental context and in that time circumstance. In order to identify which is the best solution, it is necessary to consider the positive and negative consequences of the possible choices, and then choose the option that maximises the benefits and minimises the disadvantages. But even the 'best' solution may have

S. Peppoloni and G. Di Capua, *Geoethics*,
https://doi.org/10.1007/978-3-030-98044-3_6

negative consequences, which will then have to be carefully evaluated and, finally, accepted.

From the perspective of geoethics, making technical-scientific choices under conditions of uncertainty (Albarello, 2015; Tinti et al., 2015), i.e., having incomplete knowledge of the elements on which the decision to be made will be based, inevitably means having to accept unavoidable compromises or having to manage unpleasant and risky circumstances, even on a personal level. Ovadia and Bilham (2018) report the case of a geologist employed in a mining company who was under considerable pressure from the board of directors, to the point where he had to show himself in favour of the economic feasibility of a mine in order to not risk losing his job. Another case is that of a volcanologist charged with providing the government with information on whether to evacuate the population from an area threatened by eruption (a very costly operation), who is subjected to strong political, legal, and media pressure to give certain indications in a circumstance where there are no absolute certainties, such as that of an evolving eruptive phenomenon. In general, any geologist working in an applied field, especially in the private sector, knows that their career could be jeopardised if they refuse to endorse projects on the grounds of motivated technical doubts.

Beyond personal positions, a decision on the feasibility and appropriateness of a given intervention, for example in the infrastructure or energy sector, depends not only on technical-scientific considerations, but also on multiple economic, political, social and cultural aspects. The choice will depend on assessing these factors from the perspective of and respecting the responsibilities, expectations, interests, competencies and values not only of the geoscientist, but also of the other parties involved in the intervention and especially of the community that will be affected, although these factors are often in conflict with each other.

Moreover, the impact of the intervention must also be assessed in relation to the variations of the parameters of space and time, which are fundamental in any action concerning the territory. A dam can certainly have positive effects in the short term (including flood control and the availability of water for agricultural purposes), but in the long term it can cause unpredictable and irreversible disruptions to the natural system into which it is inserted. These effects can also occur kilometres away, at the land/sea interface, where the reduction of debris reaching the coast caused by the dam can lead to increased coastal erosion and damage to existing structures. Therefore, if on one hand dams offer the possibility of producing hydroelectric energy, which is renewable and therefore 'green', on the other hand their impacts on the river ecosystem are considerable, even when their dimensions are modest, because the artificial damming of the watercourse has anyhow a disruptive effect on the delicate existing geo-ecological balances (Lerer & Scudder, 1999; Reynolds, 2011; Schmutz & Moog, 2018; Tortajada et al., 2012).

Similarly, mining may pose a potential or actual threat to the environment and health of the local communities involved, depending on how it is implemented and managed, but it undoubtedly also represents a development opportunity for economically depressed areas, as it is able to bring economic benefits, such as the creation of jobs and induced activities or the construction of infrastructure that can be used by the local population. Moreover, it is able to increase state revenues through royalties, which can be invested in welfare. Finally, the mine can also produce raw materials needed to develop low-carbon technologies for countries other than the specific country in which it is located, with undoubted collective benefits on a transnational scale.

Therefore, in any choice related to anthropic intervention on the territory, positive and negative aspects must be carefully analysed from different perspectives, assessing their impact on a local, regional, and global scale, and their effects in the short and long term. In fact, choices that do not take into account the possible evolution in time and space of the impacts on the environment and on the existing human realities may appear functional in the short term but turn out to be unfavourable in the long term, or may be advantageous within a limited range of action but be capable of triggering harmful processes even kilometres away from the place of intervention.

In addition to these aspects, it should be borne in mind that any action on the territory that is imposed on the population without the latter being involved, or at least informed, risks provoking tensions and legitimate hostile reactions, even when a reasonable alignment of values (economic, social, cultural and moral) could have been achieved through participatory discussions with all the parties involved in the intervention.

Negative examples of this are recurrent: decisions are often made on projects with a major impact on the territory (large infrastructures, underground drilling, reservoirs, landfill or nuclear waste deposits) without prior consultation with the local population and an inclusive assessment process of the social and environmental costs associated with the project. Only a dialogic collaborative action will be able to reconcile reference values and interests of the parties, bringing benefits to all those involved, maximising social and environmental benefits (Arvanitidis et al., 2017; Hostettler, 2015; Owen & Kemp, 2013; Pölzler & Ortner, 2017).

Sustainability, adaptation, prevention, and community resilience are increasingly perceived as topics of global interest, indispensable for resolving frictions and conflicts between parties in search of a functional balance (Peppoloni et al., 2019).

6.2   Facing Dilemmas: Scenarios and Uncertainties

Geoethics focuses its analysis on the role of geoscientists in the decision-making chain, since they have the greatest technical and scientific expertise on how the Earth system works, both when analysing local geological phenomena and when studying more complex global processes. Their experience is indispensable for assessing the pros and cons of any choice in the area and minimising its impact. The choice of the route of a railway line or the location of a landfill cannot fail to take into account the indispensable technical assessments of geoscientists, their expert opinion and operational proposals; otherwise, the risks and probability of failure will increase, if not the certainty of complete failure.

As Tibaldi (2013, p. 4) summarises:

The structure of the decision-making chain of technological and operational systems can be schematized as follows: 1) Monitoring-observation of ongoing events. 2) Forecasting and scenario modelling (produced, discussed and possibly shared in "peacetime"). 3) Quantitative management of observational/forecasting uncertainty (achieved with forecasts formulated in probabilistic terms). 4) Decision-making uncertainty management (use of cost-benefit models/assessments, developed and shared in advance, capable of using the output of probabilistic forecasts). 5) Decision/choice among possible interventions. 6) Real-time implementation of interventions. 7) Ex-post evaluation of the appropriateness of the interventions and the correctness of the decisions made. 8) Systematic evaluation of the quality and adequacy of the observational, modelling, decision-making and intervention system, with related evolutionary feedback on the system itself.

Within this chain of actions, geoscientists intervene at all levels. With regard to point (5), their role is that of science advisors and not of decision-makers, since the final decision will also have to take into account elements outside their specific competencies.

Geoscientists are producers of data, developers of models and possible scenarios, holders of knowledge and experience, and quantifiers of errors and probabilities. Their task is fundamental in the maturation of decisions that will be taken by others, for example by civil protection departments, who will use that knowledge, combining it with information received from other professionals (health workers, infrastructure network managers, local technicians) to outline as completely as possible a situation that may be objectively complex and require a decision under conditions of real uncertainty.

In fact, geoscientists orient the choices of others. However, even if in a certain sense this increases their responsibility beyond purely technical-scientific issues, they should not, in any case, take responsibility for decisions that depend on factors outside their competence, such as the political or economic issues involved in a given intervention and that, therefore, belong to other professional or political actors. In this case, geoscientists are required to provide all the elements (data, information, results, expert opinions) needed to make a decision that is as sustainable as possible for that social and environmental system, while respecting general reference values such as human life and the environment, health and safety, and sustainability and prevention. For example, in the case of an earthquake sequence in a high-risk seismic zone, geoscientists could be called upon to give indications on whether or not to evacuate a built-up area, even though this decision is up to the government and local decision-making bodies which have the institutional responsibility to assess the overall vulnerability and safety conditions of that territory. At the moment, geoscientists are not able to predict the occurrence of a seismic event of certain intensity in a certain time interval in deterministic terms but only in probabilistic terms and, moreover, in the current state of scientific knowledge, these indications are not usable for civil protection actions during the emergency phase.

Therefore, even though their role should be limited to informing and correctly orienting those who are really responsible for the decision-making process, nevertheless, decision-makers often expect geoscientists to provide the solution or at least to indicate what they consider to be the most appropriate among the possible decisions to be made (Bobrowsky et al., 2017).

When there are several possible choices but each of them still entails negative consequences, the geoscientist may be explicitly asked to give an expert opinion. In such cases, the first professional duty would be to make it clear that no single solution can be offered but that several hypotheses of intervention are possible, and to accompany these hypotheses with the possible scenarios that might occur so that the potential consequences of each choice are clear. In fact, knowledge of bio-geological systems should not be considered a universal and infallible datum, and one cannot think of solving an ethical dilemma solely on the basis of geo-scientific considerations or by applying simple categories such as 'right' or 'wrong', because the choice can often consist of accepting the 'lesser evil'. Very critical situations when there are strong external influences and pressures to endorse 'solutions' that are not in line with what should be chosen in that circumstance are paradoxically the most favourable opportunity to measure one's sense of responsibility (Marone & Peppoloni, 2017).

Faced with a dilemma, it could be helpful for geoscientists to consider what Gödel's theorem states (Smith, 2013), namely, that the notion of truth in a system is not definable within the system itself. In the case of geosciences, this would mean having to accept their inherent impossibility to offer solutions to real dilemmas, based solely on geo-scientific knowledge (Marone & Peppoloni, 2017). It follows that the final decisions should be left to decision-makers and technicians, who will also act on the basis of knowledge other than geo-scientific knowledge, while geoscientists are left with an ethical obligation to clarify all potential consequences associated with the different options and resulting scenarios, and to refrain from making decisions that are not within their prerogatives. In many cases, it might be useful for them to perform a cost-benefit analysis in cooperation with other professionals (Potthast, 2015; Stefanovic, 2015), considering positive and negative impacts from multiple perspectives, so that issues are also framed in social, environmental, and economic terms. But such analyses must take into account the epistemic uncertainties of the system under consideration, which might preclude the possibility of identifying and providing an optimal, or even acceptable, solution (Peppoloni & Di Capua, 2018).

Such issues are of great importance in the relations between science, politics, and society, often characterised by a lack of clarity, ambiguity, and misunderstandings, with even judicial consequences, as in the case following the 2009 L'Aquila earthquake (Amato et al., 2015; Cocco et al., 2015). Moreover, they can also lead to scientific disputes fomented by unjustifiable media competitions, as happened in the case of the 2011 Christchurch earthquake in New Zealand (Gluckman, 2014). This can have the sole effect, in the eyes of public opinion, of undermining the credibility of the scientific community and the ability to dialogue within it in order to find a position of synthesis to be proposed to society. This is very similar to what happened in several countries in the medical field, in the heated discussions on television and in the press between virologists and infectious disease specialists during the SARS-CoV-2 pandemic.

References

Albarello, D. (2015). Communicating uncertainty: Managing the inherent probabilistic character of hazard estimates. In S. Peppoloni & G. Di Capua (Eds.), *Geoethics—The role and responsibility of geoscientists* (Special Publications 419, pp. 111–116). Geological Society of London. https://doi.org/10.1144/SP419.9

Amato, A., Cerase, A., & Galadini, F. (Eds.). (2015). *Terremoti, comunicazione, diritto: riflessioni sul processo alla "Commissione Grandi Rischi"* (p. 376). FrancoAngeli. ISBN 978-8891712714.

Arvanitidis, N., Boon, J., Nurmi, P., & Di Capua, G. (2017). *White paper on responsible mining.* IAPG—International Association for Promoting Geoethics. http://www.geoethics.org/wp-respon sible-mining. Accessed 29 March 2022.

Bilham, R. (2015). M_{max}: Ethics of the maximum credible earthquake. In M. Wyss & S. Peppoloni (Eds.), *Geoethics: Ethical challenges and case studies in earth sciences* (pp. 119–140). Elsevier. https://doi.org/10.1016/B978-0-12-799935-7.00011-3

Bobrowsky, P., Cronin, V., Di Capua, G., Kieffer, S. W., & Peppoloni, S. (2017). The emerging field of geoethics. In L. C. Gundersen (Ed.), *Scientific integrity and ethics: With applications to the geosciences* (Special Publications 73, pp. 175–212). American Geophysical Union. https://doi.org/10.1002/9781119067825.ch11

Cocco, M., Cultrera, G., Amato, A., & Braun, T. (2015). The L'Aquila Trial. In S. Peppoloni & G. Di Capua (Eds.), *Geoethics—The role and responsibility of geoscientists* (Special Publications 419, pp. 43–55). Geological Society of London. https://doi.org/10.1144/SP419.13

Elster, J. (2009). Interpretation and rational choice. *Rationality and Society, 21*(1), 5–33. https://doi.org/10.1177/1043463108099347

Gluckman, P. (2014). Policy: The art of science advice to government. *Nature, 507*(7491), 163–165. https://doi.org/10.1038/507163a

Hostettler, D. (2015). Mining in indigenous regions: The case of Tampakan, Philippines. In M. Wyss & S. Peppoloni (Eds.), *Geoethics: Ethical challenges and case studies in earth sciences* (pp. 371–380). Elsevier. https://doi.org/10.1016/B978-0-12-799935-7.00030-7

Lerer, L. B., & Scudder, T. (1999). Health impacts of large dams. *Environmental Impact Assessment Review, 19*(2), 113–123. https://doi.org/10.1016/S0195-9255(98)00041-9

Lindsay, P. H., & Norman, D. A. (1972). *Human information processing: An introduction to psychology* (p. 800). Academic Press Inc. ISBN 978–0124509320.

Marone, E., & Peppoloni, S. (2017). Ethical dilemmas in geosciences. We can ask, but, can we answer? *Annals of Geophysics, 60, Fast Track 7: Geoethics at the heart of all geoscience.* https://doi.org/10.4401/ag-7445

Ovadia, D., & Bilham, N. (2018). Geoethics—What do you think? *Geoscientist—The Fellowship Magazine of the Geological Society of London, 28*(3), 7–8. https://www.geolsoc.org.uk/~/~/media/shared/documents/geoscientist/2018/Geo_APRIL2018_WR.pdf. Accessed 29 March 2022.

Owen, J. R., & Kemp, D. (2013). Social licence and mining: A critical perspective. *Resources Policy, 38*(1), 29–35. https://doi.org/10.1016/j.resourpol.2012.06.016

Peppoloni, S., Bilham, N., & Di Capua, G. (2019). Contemporary geoethics within the geosciences. In M. Bohle (Ed.), *Exploring geoethics—Ethical implications, societal contexts, and professional obligations of the geosciences* (pp. 25–70). Palgrave Pivot. https://doi.org/10.1007/978-3-030-12010-8_2

Peppoloni, S., & Di Capua, G. (2018). Ethics. In P. T. Bobrowsky & B. Marker (Eds.), *Encyclopedia of engineering geology.* Encyclopedia of Earth Sciences Series. Springer. https://doi.org/10.1007/978-3-319-12127-7_115-1

Pölzler, T., & Ortner, F. (2017). Ethical but upsetting geoscience research: A case study. *Annals of Geophysics, 60, Fast Track 7: Geoethics at the heart of all geoscience.* https://doi.org/10.4401/ag-7506

Potthast, T. (2015). Toward an inclusive geoethics—Commonalities of ethics in technology, science, business, and environment. In M. Wyss & S. Peppoloni (Eds.), *Geoethics: Ethical challenges and case studies in earth sciences* (pp. 49–56). Elsevier. https://doi.org/10.1016/B978-0-12-799935-7.00005-8

Reynolds, I. (2011). Impact of the three gorges dam. *JCCC Honors Journal, 2*(2), 3. http://schola rspace.jccc.edu/honors_journal/vol2/iss2/3. Accessed 29 March 2022.

Schmutz, S., & Moog, O. (2018). Dams: Ecological impacts and management. In S. Schmutz & J. Sendzimir (Eds.), *Riverine ecosystem management.* Aquatic Ecology, Series 8. https://doi.org/10.1007/978-3-319-73250-3_6

Simon, H. A. (1985). *Causalità, razionalità, organizzazione* (p. 416). Il Mulino. ISBN 978-8815009340.

Smith, P. (2013). *An introduction to Gödel's Theorems* (2nd ed., Cambridge Introductions to Philosophy, p. 404). Cambridge University Press. ISBN 978-1139149105. https://doi.org/10.1017/CBO 9781139149105

Stefanovic, I. L. (2015). Geoethics: Reenvisioning applied philosophy. In M. Wyss & S. Peppoloni (Eds.), *Geoethics: Ethical challenges and case studies in earth sciences* (pp. 15–23) Elsevier. https://doi.org/10.1016/B978-0-12-799935-7.00002-2

Tibaldi, S. (2013). Catena di responsabilità e catena decisionale: problemi irrisolti. Incontro-Dibattito "Cosa non funziona nella difesa dal rischio idrogeologico nel nostro paese? Analisi e rimedi", Roma, 23 marzo 2012. *Atti dei Convegni Lincei, 270*, 77–95.

Tinti, S., Armigliato, A., Pagnoni, G., & Zaniboni, F. (2015). Geoethical and social aspects of warning for low-frequency and large-impact events like Tsunamis. In M. Wyss & S. Peppoloni (Eds.), *Geoethics: Ethical challenges and case studies in earth sciences* (pp. 175–192). Elsevier. https://doi.org/10.1016/B978-0-12-799935-7.00015-0

Tortajada, C., Altinbilek, D., & Biswas, A. K. (2012). *Impacts of large dams: A global assessment* (p. XIV+410). Springer. ISBN 978-3642235702. https://doi.org/10.1007/978-3-642-23571-9

Chapter 7
The Values of Geoethics

7.1 Making Ecological Feeling Concrete

The aim of geoethics is to identify shared values on which to base strategies and operating procedures that are more responsible towards social-ecological systems, compatible with respect for the natural environment, the vocation of the territory, and the health and safety of human communities. These strategies must be scientifically defined and contextualised in time and space, i.e., they must consider different time perspectives in the analysis of expected or possible effects and take into account the diversity of social, cultural, political and economic contexts existing in the places where action is taken.

In order to better understand the responsibilities of human agents and to guide their actions, it is essential to identify reference values capable of guiding choices and behaviours within different scenarios, and on the basis of which to discriminate optimal or at least acceptable decisions from those that are less advantageous or even deleterious. These values must be rooted in awareness of the social and environmental implications of the actions of individuals and human groups and of the related responsibilities towards society, future generations and the planet.

In geoethics, responsible human action is specified and qualified through concepts such as geo-conservation, sustainability, adaptation to changes, risk prevention and geo-environmental education, which give operational concreteness to modern ecological feeling (Peppoloni et al., 2019) and which underlie the profound cultural, technological, energy and economic changes in our societies (Klein, 2011). The primary task of geoethics is to make society aware of the importance of these concepts and the need to adopt them as social reference values to ensure security and genuine progress. Indeed, it is through them that the relationship between human beings and the variety of the abiotic (geodiversity) and biotic (biodiversity) space in which they live is structured (Peppoloni & Di Capua, 2016; Peppoloni et al., 2019).

7.2 Geoheritage and Geo-conservation

Geoethics grasps the socio-cultural significance, even before the scientific significance, of geoheritage and the concept of geo-conservation and geodiversity, on which
it is possible to build new pathways of human development and be able to recover
a sense of place and strengthen the feeling of belonging to one's own spaces of
existence.

Geodiversity[1] has been defined as 'the variety of nature elements, such as
minerals, rocks, fossils, landforms and their landscapes, soils, and active geolog-
ical/geomorphological processes' that form and alter them (ProGEO, 2017) and
sustain life. Together with biodiversity, geodiversity constitutes the natural diversity
of planet Earth. 'Geoheritage comprises those elements of Earth's geodiversity that
are considered to have significant scientific, educational, cultural/aesthetic, ecological or ecosystem service values' (Woo, 2017). Geo-conservation consists of a set
of actions to ensure that the planet—the set of rocks, landscapes, waters, soils, but
also the human forms, cultures and activities that locally shape the territory—is
adequately protected from those human interventions that degrade and damage the
geosphere, atmosphere, cryosphere, hydrosphere, and biosphere, compromising its
use in the future (Bobrowsky et al., 2017; ProGEO, 2017).

Geoheritage and geodiversity visually and symbolically express the existing link
between the physical, biological, and cultural worlds. In the geoethical vision, their
conservation is fundamental, as they are irreplaceable elements of a non-renewable
social and natural 'capital'. They become points of reference to redefine the intimate
connection between human beings and Earth, thus assuming a meaning of value to
be placed at the basis of a new way of living the land.

Initiatives such as the UNESCO Global Geoparks or geotourism[2] represent their
concrete implementation as activities that can enhance the geological environment
and landscape, while also fostering a broader understanding of the sense and importance of geosciences through their learning and enjoyment (Gordon, 2018). The
UNESCO geoparks are

> single unified geographical areas in which sites and landscapes of international geological
> significance are managed with a holistic concept of protection, education and sustainable
> development. A UNESCO Global Geopark uses its geological heritage, in connection with
> all other aspects of the area's natural and cultural heritage, to enhance awareness and under-
> standing of key issues facing society, such as using of our earth's resources sustainably,
> mitigating the effects of climate change and reducing natural disasters-related risks.[3]

---

[1] The 41st session of the UNESCO General Conference established the International Geodiversity
Day on 23 November 2021. The 6th of October will be an annual worldwide celebration, raising
awareness across society about the importance of non-living nature for the well-being and prosperity
of all living beings on the planet. https://www.geodiversityday.org/. Accessed on 29 March 2022.

[2] International Congress Arouca 2011: Arouca Declaration on Geotourism. http://www.unesco.
org/new/fileadmin/MULTIMEDIA/HQ/SC/pdf/Geopark_Arouca_Declaration_EGN_2012.pdf.
Accessed 29 March 2022.

[3] http://www.unesco.org/new/en/natural-sciences/environment/earth-sciences/unesco-global-geo
parks. Accessed 29 March 2022.

Therefore, geoparks also represent the common ground on which geosciences and human/social sciences interact, offering undoubted advantages. They make multidisciplinary work and international cross-boundary cooperation tangible and effective, they produce an increase in public awareness and preparedness, they improve the quality of life of the local population by creating incentives for economic development, and, finally, they guide society towards greater responsibility for the geo- and biodiversity of nature.

7.3 Sustainability

A key concept in global change matter is sustainability, which is based on the awareness that the natural resources of the Earth system are finite. Over the past fifty years, the recognition of its value has increased in the economic and environmental fields and has become richer in its meanings (Giovannini, 2018).

In 1972, with the report 'The Limits to Growth' (Meadows et al., 1972), the Club of Rome (a non-profit association of scientists, economists, entrepreneurs and politicians) brought to the world's attention the possible consequences of continuous population growth on Earth's ecosystem and on the very survival of the human species, by means of computer simulations of complex systems governed by feedback mechanisms, produced by scientists at MIT (Massachusetts Institute of Technology) in Boston. The report pointed out that if the rate of growth of population, industrialization, pollution, food production and the exploitation of non-renewable natural resources of that time had continued unabated, within the next hundred years there would have probably been a collapse of the population, the industrial system, and the ecosystems. The same report also indicated that a condition of economic and ecological sustainability would have been only possible through a common vision of the future, able to ensure that the needs of each person were met and that everyone was guaranteed equal opportunities to realise their human potential.

In 1987, the Brundtland Commission of the United Nations introduced the concept of 'sustainable development', defining it as 'development that meets the needs of the present without compromising the ability of future generations to meet their own needs. The concept of sustainable development does imply limits – not absolute limits but limitations imposed by the present state of technology and social organization on environmental resources and by the ability of the biosphere to absorb the effects of human activities. But technology and social organization can be both managed and improved to make way for a new era of economic growth' (WCED, 1987). This definition relates the concept of sustainability to the need for natural resources and the right of all to economic and social development.

In the case of natural resources, sustainability is a concept deeply linked to human needs (Grunwald, 2015). In a broader sense, this definition invites reflection on issues of social and environmental justice, intra- and intergenerational justice, fair distribution of resources and opportunities, as well as on the concept of democracy, since it calls, albeit not explicitly, for shared governance at local and global levels, thus also introducing the concept of 'sustainability ethics' (Becker, 2012; Nurmi, 2017; Ott, 2014). In 1995, the concept of 'ecological footprint' was introduced in order to define a complex indicator to be used to assess human consumption of natural resources and energy in relation to Earth's capacity to regenerate them (Wackernagel & Rees, 1995).

Unfortunately, sustainability is a social value that is still neglected in practice, or at least underestimated in economic and social paradigms and in the decision-making processes in which these are articulated, to the point that it is reasonable to reiterate the unsustainability of current development models (Giovannini, 2018). Including sustainability systematically and simultaneously in the environmental, social, and economic perspective requires a widespread cultural change, in which politics and economics must clearly assume the responsibilities arising from their operational roles.

Affirming that there is a single living community on Earth of which humanity is an inseparable part, and that the life and future of the human species depends on Earth, is an implicit recognition that respecting Earth requires the judicious consumption of its resources, the prudent use of minerals and energy, and the care of the ecosystems that support those resources. But it is equally clear that sustainability and development need to co-exist and that it is our responsibility to explore ways of reconciling them, for example through the implementation of the concept of 'restorative sustainability', which ensures that interventions such as resource extraction produce a net benefit rather than a net detriment (Wessel, 2016). Again, the contribution of geoscience research is indispensable at several levels, not least in revealing negative consequences, contradictions, and ambiguities hidden in operational decisions (Stewart & Gill, 2017; Wyss & Peppoloni, 2015).

In fact, often the choices considered most sustainable and environmentally friendly are not so, at least in an absolute sense. This is the case with some strategies to reduce the use of fossil fuels and increase renewable energy production which, in contrast, require the use of significant amounts of non-renewable mineral resources (Nickless, 2017) and, consequently, an increase in human impact on protected natural areas and their biodiversity (Sonter et al., 2020). In addition, the extraction and processing of minerals and the related complex global supply chains, if not carefully managed, can lead to significant environmental risks and serious social damage. This is the case of 'conflict minerals' from high-risk areas or affected by armed conflicts, where human rights violations, exploitation of child labour, precarious and dangerous working conditions, illegal and unprotected mining (*artisanal mining*[4]) and other less visible impacts on local communities are widespread.

Today, the concept of sustainability is interpreted in a broader sense to include the ability of human societies to achieve or maintain certain economic, socio-political, and ecological standards. The 17 Sustainable Development Goals (SDGs) of the United Nations are a powerful effort to put a series of targets on the international political agenda, possibly quantifiable (Giovannini, 2018), to be pursued across the

---

[4] 2020 State of the Artisanal and Small-Scale Mining Sector: https://delvedatabase.org/uploads/res ources/Delve-2020-State-of-the-Sector-Report-0504.pdf. Accessed 29 March 2022.

planet by 2030, aimed at reducing economic and social inequalities between states and within them between richer and poorer areas, transition to a 'greener' economy, and preserving ecological systems. The goals identify a common matrix for action which is intended to generate a set of development pathways of societies towards a better world. The goals include the eradication of poverty (Goal 1) and world hunger (Goal 2), good health, well-being (Goal 3) and quality education for all (Goal 4), gender equality (Goal 5), clean water and sanitation (Goal 6), affordable and clean energy (Goal 7), decent work and economic growth (Goal 8), and investment in industry, innovation and infrastructure (Goal 9), reducing inequalities (Goal 10), developing sustainable cities and communities (Goal 11), responsible consumption and production (Goal 12), climate action (Goal 13), conserving the marine environment and its resources (Goal 14) and biodiversity (Goal 15), building peace, justice, and strong institutions (Goal 16), and creating partnerships to achieve the goals (Goal 17). This last objective underlines the urgent need to establish networks of national and international relations, both institutional and solidarity-based, which, acting in an agreed manner and timeframe, can contribute to the economic, social, and cultural change on a local and global scale required to achieve a human society that is universally more just, fairer, healthier, safer, economically and politically developed, and ecologically responsible.

From the point of view of geoethics, the 17 SDGs constitute for humanity a perfect common framework of ideal references and concrete actions to develop. Moreover, they are a great opportunity to highlight the fundamental contribution that geosciences can offer to society (Gill, 2016; Gill & Smith, 2021). For states, they can be the starting point for achieving planetary governance that overcomes differences in economic and social systems in view of a common goal.

However, careful analysis of the 17 SDGs highlights some possible contradictions that may make this reference framework ambiguous, in particular the fact that no priority is established in the actions to be taken and that the concept of sustainability is closely linked to that of economic development. This link is certainly feasible but not within the current systemic framework, which uses models of economic and social development and ways of calculating the value of the final goods and services produced by states based on Gross Domestic Product (GDP) that are completely incompatible with sustainability. In fact, there are many voices that denounce the capitalist system, not without prejudice and ideological aversions, as the primary cause of the planetary ecological crisis (Bonneuil & Fressoz, 2013; Lewis & Maslin, 2018), although the actual roots of the crisis are perhaps to be sought even deeper, in Western culture, as White Jr. already stated (1967).

It should be added that the concept of sustainable development initially originated with reference to natural resources. In the most widespread narrative and common perception, sustainability is mainly understood in terms of a sustainable economy or actions aimed at environmental sustainability. And even in the latter case, its meaning refers to the fight against the degradation of the natural environment in relation to its economic significance, and not rather, as it should also be, to the preservation of those physical, biological, and aesthetic qualities of natural environments, the degradation of which affects people's emotional and health aspects. Moreover, speaking about 'environmental sustainability' is to recall the need for human actions that ensure the sustainability of nature, despite the fact that nature already possesses an intrinsic sustainability in itself and the ability to adapt to changing conditions. However, what could irreversibly change, due to anthropogenic or natural causes, and thus needs responsible human action, is a natural environment capable of ensuring sustainability for human life and other living species.

The 17 SDGs defined in the 2030 UN Agenda, whose value can be compared to that of the UN Declaration of Human Rights, and the tension towards the transition to a more sustainable economy are certainly epoch-making facts that geoethical thought embraces, emphasizes, and promotes. However, rather than talking about 'sustainable development', geoethics proposes the concept of 'responsible human development', which also includes economic development as a transition to an essentially circular and reuse economy with a reduced ecological impact. 'Responsible human development' is a broader concept than simply 'sustainable development', as it assigns humanity the task and responsibility of making society sustainable, so that it is fair, equitable, supportive, inclusive, educated, participatory and ecologically oriented. Aiming for responsible human development means increasing the overall social, economic, cultural and ecological awareness and responsibility of human agents, according to the possibilities of each one, in order to implement a true progress of civilisation.

7.4   Adaptation

Human adaptation refers to the ability of a social group to modify its organisation, modes of production and consumption, interests, goals, network of external relations and the ways in which it interacts with its environment in response to change. Natural systems modify irreversibly because of their interconnectedness and complexity, due to the non-linear dynamics that govern them and which do not allow full restoration of previous conditions. Human communities must develop strategies and actions to adapt to natural and anthropogenic changes to ensure their survival and, in pursuing this common advantage, they strengthen their internal social bonds.

At the planetary level, global warming is a major concern today, with more than 95% of scientific studies now confirming its anthropogenic origin (Oreskes, 2014). For Lynas et al. (2021), this consensus is even greater than 99%. Since the formation of the planet, climate and average temperatures on Earth have changed for different reasons (for changes in average temperatures over the last 66 million years, see Westerhold et al., 2020). These include the effect of the movement of lithospheric plates that changes the size and arrangement of continental masses relative to the oceans; gaseous and pyroclastic emissions from volcanoes; changes of Earth's orbit and the inclination of its axis; changes in the amount of oxygen emitted by plants; and changes in the extent of forested areas capable of storing carbon dioxide (CO_2). Thanks to a small amount of CO_2, the greenhouse effect ensures that Earth's surface maintains a temperature compatible with the development and spread of life. In fact, carbon dioxide is able to effectively capture the heat emitted by Earth and prevent it from dissipating rapidly into space. The problem, however, is that human activities are releasing quantities of CO_2 into the atmosphere that are even thirty times higher than natural ones, causing the heat to be trapped between Earth's surface and the atmosphere for a long time before it finally dissipates into space, leading to a rise in global average temperature (Ripple et al., 2020, 2021). Many scientists predict that in a few decades this trend will lead to average temperature values reached on Earth millions of years ago, when the climate was considerably warmer, the ice caps were missing, and sea levels were several tens of meters higher than today.

Obviously, climate and temperature changes are not only a problem of our historical epoch. Even in the past, humanity has had to cope with the heavy effects of climate and temperature variations, trying to adapt to changed environmental conditions and to defend itself from the consequent less probable events (commonly called 'extremes') or even facing migrations (Brooke, 2014; D'Arrigo et al., 2020; Gill, 2001). But changes that occurred in historical times generally had a regional extension, whereas current changes are affecting the entire planet, with increases in average temperatures not only in the atmosphere but also in ocean waters. The responsibility of modern humans for accelerating the process of global warming, particularly after World War II, is increasingly evident (Steffen et al., 2015), whereas just a few decades ago this awareness was not so obvious. After all, the unpredictable (or unlikely) event is always possible in nature, especially when human beings act without foreseeing the consequences of their actions.

The scientific debate on climate change is very heated because of the enormous repercussions that the induced transformations could have on our lives but, fortunately, there are many more tools available than in the past to study, understand, and find the most suitable strategies to avoid reaching the threshold limit for our survival on the planet. Nowadays science is able to set up increasingly reliable climate and temperature scenarios for the future (Mearns et al., 2001) using sophisticated numerical models that provide predictive images of the conditions in which our planet may be in 10, 50, 100 years and more (Collins et al., 2013; Schwalm et al., 2020). Models are a valuable aid in predicting possible adverse effects and developing preventive actions. However, this is not enough. Models need to be followed by policy measures that mitigate the negative consequences of changing environmental conditions on social and economic balances, both at local and global scales.

With reference to biological systems, adaptation—that evolutionary process by which living beings morphologically and physiologically transform themselves in order to continue living in modified environmental conditions—becomes fundamental. Adaptation can determine not only the fate of individuals and populations but the success or failure of a species in evolutionary terms. The human species has always had to adapt to environmental changes, initially in biological terms but later also through cultural changes (Foley et al., 2013). Adaptation to environmental change is about trying to reduce the vulnerability of social-ecological systems and to mitigate and compensate for the effects of global warming, albeit with 'barriers, limits and costs, but these are not fully understood' (from Climate Change Report

2007).[5] Adapting and mitigating can mean decreasing the vulnerability of the system or community by reducing its sensitivity to change and also increasing its resilience (Adger et al., 2005), i.e., its ability to adapt. But the need to adapt can also create new opportunities for development (Betsill, 2001; Conway & Schipper, 2011), for example when investments in innovative research and technologies are planned, or vulnerability to other hazards is also reduced, or the development of new and sustainable economic pathways is encouraged. In addition to these benefits, an increased awareness of our global interdependence and, thus, the need for common responses is in itself a positive outcome, which can encourage citizens and governments to be more proactive and cooperative.

In this respect, the 2015 Paris Agreement (COP21),[6] signed by 195 nations despite its implementation limitations, demonstrates the growing political will of the international community to act in this direction. The agreement sets shared goals to limit CO_2 emissions and provides a reference framework for the actions and investments needed to ensure a low-emission future that is adaptive, resilient and sustainable.

Many expectations and concerns accompanied the 2021 COP26. Part of the international public opinion would have wished that event to become a point of arrival of a long process started in 1992: the global efforts of states to start an effective trend of reducing greenhouse gases to fit the goal of 1.5 °C increase of the global average temperature above pre-industrial levels should have led to resolute and ambitious decisions. Certainly the economic consequences of the SARS-CoV-2 pandemic and the emergence of some geopolitical frictions have influenced the outcome of COP26, weakening the negotiating skills of some nations and slowing down the process of definitive abandonment of fossil fuels by state parties. Between lights and shadows, the most significant results are the following:

- The threshold of 1.5 °C has been accepted by all parties as the goal to be pursued to keep global warming within an acceptable limit, in order to avoid catastrophic climatic consequences for humanity. To achieve this goal, global greenhouse gas emissions must be reduced by at least 45% by 2030. The reduction commitments will be checked annually to make any necessary changes to adapt to the evolution of events. This seems to be one of the best results of COP26, since at least it binds the states to an annual control of the environmental conditions of the planet and to intervene, if necessary, with more decisive and stringent actions.
- More than 100 states have pledged to end deforestation by 2030.
- The United States and the European Union will work to reduce methane gas emissions both from extraction and from distribution infrastructures (gas pipelines).
- Efforts will be intensified for 'phasing down' (as requested by India) and no longer for 'phasing out' (as initially provided in the final draft of the closing document of

---

[5] "The Impacts of Climate Change, Adaptation and Vulnerability", summary for policy makers, in https://www.ipcc.ch/site/assets/uploads/2018/02/ar4-wg2-spm-1.pdf (p. 19). Accessed 29 March 2022.

[6] https://unfccc.int/process-and-meetings/the-paris-agreement/the-paris-agreement. Accessed 29 March 2022.

the conference) of coal without capture of CO_2 and subsidies for fossil fuels. This decision has created much disappointment within the international community, especially among environmental movements. In any case, for the first time states agree on the danger of coal more than other fossil fuels.

- The richest countries have renewed their commitment to create a fund of USD 100 billion per year to contribute to energy transition from fossil fuels to renewables of the poorest states or states most exposed to climate change. This fund is expected to be increased by USD 200 billion per year from 2025.
- Some rich nations have pledged USD 1.7 billion in aid for policies to support Indigenous peoples.
- The United States and China have signed a bilateral agreement to initiate climate cooperation. More than an agreement of substance, this is a sign of hope that comes from the two main superpowers of the planet.
- States are calling for finance, banks and other financial institutions to commit to supporting governments and the economy in energy transition.

Leaving the failure-success duality, we can ultimately consider COP26 as a further step in the right direction, even if still, perhaps, too timid, especially considering the absence of references in official documents to the historical responsibilities for the pollution of the planet of the richest countries and to the primary causes of the ecological crisis, that is, the dominant development models.

It must also be taken into account that some developing nations, more than Western nations, are reluctant to implement ambitious policies to reduce greenhouse gas emissions and are unwilling to make certain decisions for legitimate reasons related to the development of their emerging economies (India) or to the energy supply of their massive production and distribution facilities (China).

But we cannot fail to notice how long it has taken to reach any agreement of governments to fight global warming and begin to really work on the physical and chemical processes that are altering the climate. In fact, a long process started in 1992 with the United Nations Framework Convention on Climate Change (UNFCCC). Significantly, the state parties agreed that 'change in the earth's climate and its adverse effects are a common concern of humankind' and 'that human activities have been substantially increasing the atmospheric concentrations of greenhouse gases, that these increases enhance the natural greenhouse effect, and that this will result on average in an additional warming of the earth's surface and atmosphere and may adversely affect natural ecosystems and humankind...'.[7]

Then, the process continued with the Kyoto Protocol, the international treaty signed in 1997 that commits 'industrialized countries and economies in transition to limit and reduce greenhouse gases (GHG) emissions in accordance with agreed individual targets... (and) sets binding emission reduction targets for 37 industrialized countries and economies in transition and the European Union..'[8] This was an

---

[7] https://unfccc.int/files/essential_background/background_publications_htmlpdf/application/pdf/conveng.pdf. Accessed 29 March 2022.

[8] https://unfccc.int/kyoto_protocol. Accessed 29 March 2022.

important step in adopting concerted global actions, finally followed by the Conferences of the Parties (COP) in order to enlarge the number of parties actively involved in active policies to contrast global warming, looking to the problems also in wider perspective capable to consider also social inequalities as an effect of environmental problems.

The results of COP26 more and more demonstrate that the ecological crisis must be faced by establishing an authoritative and permanent supranational governance capable of synthesizing the various national instances with continuity, thanks also to a precise mandate from governments. In fact, the negotiations and agreements between individual states within the framework of the COPs do not seem able to satisfy the need for determination that the decisions of the next decades require.

In this scenario, the scarce or even absent consideration for the contribution that geosciences can make to governments in planning policies and implementing best practices in the field of sustainability is rather embarrassing. The geosciences community will have to question itself deeply about the reasons and responsibilities of this absence.

A significant precedent of international cooperation to face a global environmental emergency is set by the Montreal Protocol signed in 1987,[9] which established effective international policies to combat ozone depletion and which, in a relatively short time, succeeded in reversing the downward trend of this fundamental atmospheric gas in the 2000s.

It is worth mentioning that the Green Deal,[10] approved by the European parliament on 15 January 2020, is more than a positive signal to radically change the current social and economic policies of a large part of the rich and unsustainable Western world. The Green Deal aims to make the economy of the European Union sustainable by promoting the efficient use of resources through a shift to a clean and circular economy, the restoration of biodiversity, and the reduction of pollution.

7.5   Prevention

Risk is the possibility that a natural phenomenon of a given intensity may cause damage to the population, residential settlements, productive activities and infrastructures within a particular area, in a predefined time interval. It can be defined as the symbolic product of hazard (P), vulnerability (V) and exposure (E):

$$R = P \times V \times E$$

Hazard is the probability of occurrence of a phenomenon of predefined intensity in a given area and time interval. Hazard is an intrinsic element of the territory, a

---

[9] https://ozone.unep.org/treaties/montreal-protocol. Accessed 29 March 2022.

[10] https://ec.europa.eu/info/strategy/priorities-2019-2024/european-green-deal_en. Accessed 29 March 2022.

function of its geological, morphological, and climatic characteristics. A distinction can be made between a natural hazard, e.g., linked to the presence of landslides or the occurrence of floods, earthquakes and volcanic eruptions, and a hazard induced or increased by anthropogenic activities. These hazards include those caused by river barrages, storage of materials, soil erosion accelerated by agricultural activities, instability of artificial slopes that have been wrongly dimensioned or of natural slopes that have been deforested or excavated, as well as from the rupture of anthropogenic embankments, waste disposal, surface and groundwater pollution. Anthropogenic global warming increases the hazard of numerous natural phenomena and, ultimately, the risk to which society is exposed.

Vulnerability is the capacity of an element or group of elements (buildings, bridges, electrical and railway networks) to resist a given natural phenomenon of a given intensity. It is a function of the intrinsic characteristics of the element considered which, in turn, depend on the intensity of the reference phenomenon. The vulnerability of the structures can be modified through constructive measures. There is also a social vulnerability linked to the intrinsic capacity of a community to resist natural phenomena that affect it, which may depend on factors of different nature (economic, cultural, historical, social, psychological). Social vulnerability is linked to resilience, i.e., the ability of the community to react to a disaster and to recover.

Finally, exposure is the number or value of items at risk in a given area and is quantified in terms of human life, economic or historical-artistic value.

Risk implies the presence in the territory of 'elements' that can be damaged (population, residential settlements, productive activities, infrastructures, cultural heritage). To concretely assess the risk, it is not enough to know the hazard, but it is also necessary to carefully estimate the value exposed, i.e., the assets present in the territory which may be affected by the event and their vulnerability.

Therefore, it is against risk that we can defend ourselves; it is by acting on the risk that prevention takes place, being aware that the risk cannot be completely eliminated but can be reduced below a threshold deemed acceptable by society (Peppoloni, 2014). In this regard, in its application to natural hazards, the geoethical thought finds correspondence in what is expressed by Giuseppe Grandori (1921–2011), a prominent figure in Italian earthquake engineering who, more than thirty years ago, stated, 'Defending oneself against earthquakes means reducing the consequences of earthquakes (victims and material damages) below a limit that society considers acceptable, taking into account the costs that a further reduction of this limit would entail' (Grandori, 1987, p. 7). The apparent cynicism of this statement disappears once one recognises the proactive scientific intentionality and the great operational and cultural value. Defending oneself against risks is the result of a path of knowledge that takes shape in a social contract and in which the dignity of human reason is an irreplaceable instrument at the service of the common good.

Prevention refers to a set of activities and tools aimed, in some cases, at preventing the occurrence of events that are dangerous to people's safety and, in others, at preventing the damage resulting from exposure to a risk. In terms of risk management, prevention strategies can be considered as being aimed at interrupting the path that links a risky event to its possible causes (proactive controls) or the event to its possible consequences (reactive controls) (Peppoloni et al., 2019).

Preventive actions can develop over long periods of time, requiring careful planning of human and economic resources, but they are the best way to increase the resilience of communities. A resilient community will be able to anticipate, avoid and/or respond to a disruptive event by restoring the material, and also cultural and spiritual, conditions that existed prior to that event and to prepare for and respond to future events more effectively.

The development of a culture centred on prevention requires providing communities with correct information and basic scientific knowledge, an effective decision-making chain capable of assessing and establishing reasonable risk thresholds, and the adoption of strategies that reduce the chances of a natural or anthropogenic event turning into a disaster.

The achievement of a fundamental objective such as risk reduction through prevention will depend on multidisciplinary research, the identification of areas at risk, the development of early warning systems and networks for the continuous monitoring of phenomena and exposed objects, the securing of the building stock or the relocation of buildings, plants, and infrastructures that are particularly at risk, the development of regulatory instruments, effective coordination between the many parties responsible for risk defence, and, finally, the organisation of information and training campaigns aimed at citizens. All these activities represent the implementation of a preventive strategy that must be at the basis of any territorial management policy.

Awareness-raising and education activities for the population represent one of the pivots on which to base a new relationship between institutions, citizens, science and the territory, and require the preventive commitment and partnership of the entire society. As clearly stated in the Guiding Principles of the Sendai Framework for Disaster Risk Reduction 2015–2030,[11] states have the main role in disaster risk prevention and reduction, and this responsibility must be shared with other components of society, including local authorities and the private sector. Prevention involves both the entire body of society and the individual, each with their own share of responsibility for looking after the common interest. There are aspects that the state must necessarily take on through structures such as civil protection departments or research bodies and universities, but there is also an individual ethical duty to improve one's own preparation and be ready to cooperate, a duty which, in addition to having value in itself and for themselves, has the advantage of producing positive effects on the entire community. The lack of preparation does not only concern politicians or local government technicians but all citizens, who too often delegate their own safety to the responsibility of others or even to fate when, instead, defending themselves against natural risks is the civic duty of each individual towards the community.

It is also true that citizens are usually considered passive actors in risk scenarios or land management decisions, while actually they can (and should) play an active role. The involvement of populations living in hazardous areas in educational and risk

---

[11] https://www.undrr.org/implementing-sendai-framework/what-sendai-framework. Accessed 29 March 2022.

communication activities can contribute to increased community resilience (Ickert & Stewart, 2016; Stewart et al., 2017), although there are some limitations in their effectiveness, well highlighted by Wachinger et al. (2013). In fact, not always an adequate perception of risk improves personal preparedness leading to the development of virtuous behaviours that increase one's own safety.

But is risk something that is adequately perceived? How many citizens are really aware that they are risking their lives in certain circumstances? How many have an idea of the vulnerability of their homes to floods, landslides, earthquakes, or know the safest places in their homes and towns?

In this regard, the case of Italy is significant: it is a geologically young land, in constant evolution and therefore fragile. Its difficult physical environment has often been compounded over time by carelessness, negligence, and even wicked human interventions that have further increased exposure to risk. Moreover, this situation has not been accompanied by a concomitant increase in risk perception by the population, which is consequently unable to fully understand the importance of demanding that governments develop defence and prevention policies.

Prevention is, above all, an ethical duty that we must responsibly assume out of respect for our own humanity, without neglecting in any way the economic dimension of the issue (Guidoboni & Valensise, 2011), including the need to also increase resilience to reduce the weight of the state's economic intervention to restore, as far as possible, the status quo existing before the event. The architect Pirro Ligorio (1513–1583) stated this as early as the sixteenth century in his book *Libro di Diversi Terremoti* (Guidoboni, 2005), when he reiterated that earthquakes are not obscure and ineluctable accidents, but phenomena within the reach of human reason, and that trying to achieve housing safety is a necessity and a duty of the human intellect. He clearly refers to the responsibility of human beings, who turn earthquakes into disasters (but the same applies to any other natural phenomenon), when they culpably fail to do anything that is in their rational possibilities to defend human lives, property, and activities. However, while science, in its historical evolutionary dimension, has always learned from earthquakes, floods, eruptions and other events of the past, questioning its theories and forecasting models on the basis of direct observation of what happened, modern society, and politics in particular, seem to forget too quickly the lessons of the present, postponing the adoption of long-term intervention strategies to the distant future. As a result, the constant references to necessary preventive action remain confined to speeches, misused expressions, and unfulfilled promises.

Certainly, a flood, an eruption, an earthquake will recur where geological conditions 'favourable' to their occurrence remain; it is only a matter of time. Science has been forcefully repeating this for decades. But even certain forms of popular wisdom, born of a more authentic and observational relationship with one's own territory, ensured that, in the past, people did not build in areas where unfavourable local ground conditions existed. On the contrary, modern society has restricted the temporal perspective of its action to the short term, busy pursuing contingent problems. For this reason, the 'culture of emergency' that dominates most of human societies no longer appears to be the cause but proves to be the effect of this inability to think about the future. And, in the defence against natural risks, the inability to rationally prefigure a possible future leads human beings to a constant passive attitude towards phenomena that have a return period of up to decades. In this cultural context, prevention remains an empty and worthless word.

This state of inaction is also fuelled by the ease with which people tend to lose memory of past disasters. In the case of earthquakes, the most energetic events may have very long return periods, of several tens if not hundreds of years. Such return periods exceed the lifespan of a human being, so much so that, after a seismic event, it takes only a few years to forget it. Time dilutes the memory of the event and banishes fear. And, as that memory fades, the need to pay appropriate attention to the use of construction practices in those particularly risky areas of a territory also fades from human memory. Thinking about the possibility of an earthquake with a return period of hundreds of years is something that goes beyond common experience. However, memory is an indispensable element for entering the temporal dimension of natural phenomena and understanding them. If what happened in the past is not forgotten, it will be possible to work with greater conviction to prevent what may happen in the future. And in any case, this is not enough.

Why is so difficult to implement prevention? Why is it so difficult in the present to think about the future of the territory we live in? In a perspective confined to the present, people are led to think, not without reason, that the basis of the chronic delay in launching widespread prevention policies is exclusively the difficulty in finding adequate economic resources, bureaucratic inefficiency, and short-sightedness of the ruling classes. But, alongside these reasons, there is also and above all a cultural problem.

Deferring the issue of prevention to politics alone may, to some extent, relieve citizens of the responsibility of taking their future into their own hands, remaining in a wait-and-see attitude towards state intervention. But this certainly does not facilitate the development of a knowledge-based society.

Many human societies seem to lack recognition of the values on which prevention is based. In fact, prevention activities do not only bring economic benefits but also respond rationally and responsibly to that right to safety that everyone must ethically pursue for themselves and for the community to which they belong. In order to build a real and widespread culture of prevention, it is necessary to recover the idea of the territory as a common good, a resource and a collective advantage, an element in which individual interests and those of the entire community converge. To recognise the value of that asset for oneself and for others, to respect it, defend it, preserve it and pass it on to future generations, one must first have rebuilt solid social relations within the community, based on shared and implemented ethical principles. The close relationship between geological instability and social instability dramatically underlines a collective inattention to the territory. This is not simply the place where one is accidentally born or lives but the physical support of one's activities, a precious historical, cultural, emotional and economic resource, and also, and above all, one of the founding values of human identity. It is only by rediscovering the identity value of the territories that it is possible to initiate a cultural change that can lead everyone to understand the advantages of pursuing policies to prevent natural risks, enhance the value of specific historical and artistic features, protect bio- and geodiversity and develop new economic models.

Knowledge of the value of the territory is the first building block on which to build a new social awareness, guiding citizens towards a more responsible way of thinking about and managing their living space.

7.6 Geo-environmental Education

The advancement of knowledge in the geosciences has been fundamental to the development of human thoughts and modern cultures. The theories developed, models constructed, and results produced by geoscientific research have provided better living conditions for many human societies, but they have also been the subject of philosophical reflections and incorporated as fundamental parts of human cultures. Geosciences, through their discoveries and visions, methods, goals, reference values and ways of thinking about nature, are not only a set of technical-scientific knowledge useful for solving complex problems of planetary management but also a cultural support that should accompany the practical response to those problems (Peppoloni & Di Capua, 2012). Science is a fundamental aspect of culture (Bly, 2010; Giorello & Guzzardi, 2010; Peppoloni, 2012) and geosciences make culture, contributing to the construction of that constellation of ideas and concepts that can be used to understand the world (Peppoloni & Di Capua, 2012; Raab & Frodeman, 2002; Seddon, 1996).

Few scientific disciplines are as capable of bringing humankind closer to the meaning and depth of time as when we study fossils or the geological eras of a history of the earth that has lasted for billions of years.

Through the geosciences (e.g., palaeontology) the human mind becomes accustomed to exploring counter-intuitive temporal spaces and losing itself in images of worlds far removed in time. The age of the earth (4.56 billion years) has been established by geochronological dating based on the radioactive decay of chemical elements such as uranium. It is such a vast period of time that it is difficult to conceive without an effort of the imagination. Similarly, it is not immediate for the human mind to think of an atmosphere on Earth without oxygen, or seas without life forms, or lands dominated by dinosaurs. Geology allows us to read the sequence of events that produced a rock by studying the minerals and fossils that make it up and by tracing the environments in which it was formed. Geology is able to describe the enormous distances covered by continents, the continuous collision and fragmentation of lithospheric plates, the abyssal depths of the oceans, and the constant forming and disintegration of rocks under the incessant action of endogenous and exogenous agents. In this way, we are led to imagine boundless spaces compared to those in which we usually move, or completely inaccessible like the depths of Earth.

But geosciences are also able to look into the future of the planet through predictive models, hypothesise the movements of a continent in a time interval of a few million years, and assess how the profile of the land will change as a result of the rise in the average sea level due to the melting of glaciers on land. Finally, geosciences are able to make 'real' even that which is not directly observable, such as the source of an earthquake or a magnetic field. These are objects that can only be thought of but their effects are tangible, and they can be perceived and measured with special instruments that certify their occurrence or existence.

Geo-environmental education enables geoscientists to transfer knowledge about the forms, processes, and products of natural or human-induced dynamics, past and present, providing their projection into the future. Proposed to the general public, geo-environmental education enables people to confront the categories of time and space necessary to understand and orient themselves in reality. Scientific concepts and theories such as deep time, evolutionism and plate tectonics are fundamental keys to interpreting the universe, to thinking about reality and building one's self in relation to it as well.

For this reason, geo-environmental education is an activity of great ethical value, in which the notions of true, fair, appropriate, and acceptable can be grasped in their meaning applied to social and environmental benefit. Furthermore, it stimulates systemic and critical thinking, trains the attitude that enables us to grasp the interdependence of elements, the complexity of systems, and the reversibility of thinking and observing.

Through appropriate geo-environmental education, we can delve into the images that shape the physical reality of our perceptual space. The geo-scientific culture makes us aware that Earth is a living 'entity', in continuous spatial and temporal evolution, characterised by transformations so fast or so slow that they cannot be perceived by the human mind and made up of interconnected and constantly interacting elements (Doglioni & Peppoloni, 2016).

Unfortunately, geoscientific literacy is sorely lacking in our societies, and this serious deficiency means that we are not fully able to follow the debate on environmental issues. For the same reason, most politicians, journalists or opinion makers often lack adequate knowledge of the causes and effects of environmental issues, fueling confusion and disorientation among citizens. Moreover, ignorance fosters the spread of pseudo-science, fatalistic, and conspiratorial attitudes when, instead, good preparation of citizens would improve the quality of their participation in decision-making processes on environmental, energy, infrastructure, risk defence and land management issues, a fundamental activity in any democratic society (Lacreu, 2017).

The promotion of geosciences in society through geo-education involves the introduction of innovative teaching methods and tools (Orion, 2019) aimed at developing the observational and analytical skills of students and citizens (Peppoloni et al., 2019). Geoscientists involved in geo-education exemplify the practice of geosciences as a precise geoethical duty to society, as they undertake enabling others to understand 'that we are unavoidably participative beings in Earth's systems - that our behavior is constantly affecting and being affected by everything natural and human, in a dynamic relationship' (Orion, 2019, p. 1).

Geo-environmental literacy is crucial for a peaceful coexistence with the environment, for changing the way people perceive their relationship with the Earth system, for fostering conscious participation in debates and actions aimed at solving daily problems and global challenges, as well as providing the basis for training future geoscientists, whose skills will be essential in meeting those challenges over time.

References

Adger, W. N., Arnell, N. W., & Tompkins, E. L. (2005). Successful adaptation to climate change across scales. *Global Environmental Change, 15*(2), 77–86. https://doi.org/10.1016/j.gloenvcha.2004.12.005

Becker, C. U. (2012). *Sustainability ethics and sustainability research* (p. XIV+14). Springer. ISBN 978-9400722859. https://doi.org/10.1007/978-94-007-2285-9

Betsill, M. M. (2001). Mitigating climate change in US cities: Opportunities and obstacles. *Local Environment, 6*(4), 393–406. https://doi.org/10.1080/13549830120091699

Bly, A. (Ed.). (2010). *Science is culture—Conversations at the new intersection of science + society* (p. 368). HarperPerennial. ISBN 978-0061836541.

Bobrowsky, P., Cronin, V., Di Capua, G., Kieffer, S. W., & Peppoloni, S. (2017). The emerging field of geoethics. In L. C. Gundersen (Ed.), *Scientific integrity and ethics: With applications to the geosciences* (Special Publications 73, pp. 175–212). American Geophysical Union. https://doi.org/10.1002/9781119067825.ch11

Bonneuil, C., & Fressoz, J.-B. (2013). *L'Evénement Anthropocène - La Terre, l'histoire et nous* (p. 320). Seuil. ISBN 978-2021135008.

Brooke, J. L. (2014). *Climate change and the course of global history: A rough journey* (p. 654). Cambridge University Press. SBN 978-0521871648

Collins, M., Knutti, R., Arblaster, J., Dufresne, J.-L., Fichefet, T., Friedlingstein, P., Gao, X., Gutowski, W. J., Johns, T., Krinner, G., Shongwe, M., Tebaldi, C., Weaver, A. J., & Wehner, M. (2013). Long-term climate change: Projections, commitments and irreversibility. In T. F. Stocker, D. Qin, G.-K. Plattner, & M. Tignor, S. K. Allen, J. Boschung, A. Nauels, Y. Xia, V. Bex & P. M. Midgle (Eds.), *Climate change 2013: The physical science basis*. Contribution of Working Group I to the Fifth Assessment Report of the Intergovernmental Panel on Climate Change. Cambridge University Press. https://www.ipcc.ch/site/assets/uploads/2018/02/WG1AR5_Chapter12_FINAL.pdf. Accessed 29 March 2022.

Conway, D., & Schipper, E. L. F. (2011). Adaptation to climate change in Africa: Challenges and opportunities identified from Ethiopia. *Global Environmental Change, 21*, 227–237. https://doi.org/10.1016/j.gloenvcha.2010.07.013

D'Arrigo, R., Klinger, P., Newfield, T., Rydval, M., & Wilson, R. (2020). Complexity in crisis: The volcanic cold pulse of the 1690s and the consequences of Scotland's failure to cope. *Journal of Volcanology and Geothermal Research, 389*, 106746. https://doi.org/10.1016/j.jvolgeores.2019.106746

Doglioni, C., & Peppoloni, S. (2016). *Pianeta Terra: una storia non finita. Il Mulino* (p. 160). Bologna. ISBN 978-8815263766.

Foley, S. F., Gronenborn, D., Andreae, M. O., Kadereit, J. W., Esper, J., Scholz, D., Pöschl, U, Jacob, D. E., Schöne, B. R., Schreg, R., Vött, A., Jordan, D., Lelieveld, J., Weller, C. G., Alt, K. W., Gaudzinski-Windheuser, S., Bruhn, K.-C., Tost, H., Sirocko, F., & Crutzen, P. J. (2013). The Palaeoanthropocene—The beginnings of anthropogenic environmental change. *Anthropocene, 3*, 83–88. https://doi.org/10.1016/j.ancene.2013.11.002

Gill, J. C. (2016). Geology and the sustainable development goals. *Episodes, 40*(1), 70–76. https://doi.org/10.18814/epiiugs/2017/v40i1/017010

Gill, J. C., & Smith, M. (Eds.). (2021). *Geosciences and the sustainable development goals* (p. XXXIII+474). Springer. ISBN 978-3030388157. https://doi.org/10.1007/978-3-030-38815-7

Gill, R. B. (2001). *The Great Maya Droughts—Water, life, and death* (p. 484). University of New Mexico Press. ISBN 978-0826327741.

Giorello, G., & Guzzardi, L. (2010). Ricerca scientifica e libertà politica. In A. Olmi (a cura di), *L'eredità dell'Occidente – Cristianesimo* (pp. 129–143). Europa, nuovi mondi.

Giovannini, E. (2018). *L'utopia sostenibile* (p. 160). Editori Laterza. ISBN 978-8858130766.

Gordon, J. E. (2018). Geoheritage, geotourism and the cultural landscape: Enhancing the visitor experience and promoting geoconservation. *Geosciences, 8*, 136. https://doi.org/10.3390/geosci ences8040136

Grandori, G. (1987). Introduzione. In D. Benedetti, A. Castellani, C. Gavarini, & G. Grandori (a cura di), *Ingegneria Sismica, Quaderni de "La Ricerca Scientifica"* (n. 114, Vol. 6). Consiglio Nazionale delle Ricerche. ISSN 0556-9664.

Grunwald, A. (2015). The imperative of sustainable development: Elements of an ethics of using georesources responsibly. In M. Wyss & S. Peppoloni (Eds.), *Geoethics: Ethical challenges and case studies in earth sciences* (pp. 25–35). Elsevier. https://doi.org/10.1016/B978-0-12-799935-7.00003-4

Guidoboni, E. (a cura di). (2005). Pirro Ligorio, Libro di Diversi Terremoti. In *Collana "Edizione Nazionale delle Opere di Pirro Ligorio", Torino* (Vol. 28, p. 296). De Luca Editori D'Arte. ISBN 978-8880167464.

Guidoboni, E., & Valensise, G. (2011). *Il peso economico e sociale dei disastri sismici in Italia negli ultimi 150 anni* (p. 552). Bononia University Press. ISBN 978-8873956839.

Ickert, J., & Stewart, I. S. (2016). Earthquake risk communication as dialogue—Insights from a workshop in Istanbul's urban renewal neighbourhoods. *Natural Hazards and Earth System Sciences, 16*, 1157–1173. https://doi.org/10.5194/nhess-16-1157-2016

Klein, R. J. T. (2011). Adaptation to climate change: More than technology. In *Climate: Global Change and Local Adaptation. NATO Science for Peace and Security Series C: Environmental Security* (pp. 157–168). Springer. https://doi.org/10.1007/978-94-007-1770-1_9

Lacreu, H. L. (2017). The social sense of geological literacy. *Annals of Geophysics, 60, Fast Track 7: Geoethics at the heart of all geoscience.* https://doi.org/10.4401/ag-7558

Lewis, S. L., & Maslin, M. A. (2018). *The human planet: How we created the Anthropocene* (p. 480). Pelican. ISBN 978-0241280881.

Lynas, M., Houlton, B. Z., & Perry, S. (2021). Greater than 99% consensus on human caused climate change in the peer-reviewed scientific literature. *Environmental Research Letters, 16*, 114005. https://doi.org/10.1088/1748-9326/ac2966

Meadows, D. H., Meadows, D. L., Randers, J., & Behrens III, W. W. (1972). *The limits to growth— A Report for the Club of Rome's project on the predicament of mankind* (p. 205). Potomac Associates—Universe Booksk. ISBN 0-87663-165-0. http://www.donellameadows.org/wp-con tent/userfiles/Limits-to-Growth-digital-scan-version.pdf. Accessed 29 March 2022.

Mearns, L. O., Hulme, M., Carter, T. R., et al. (2001). Climate scenario development. In *Climate Change 2001: The Scientific Basis*. Contribution of Working Group I to the Third Assessment Report of the Intergovernmental Panel on Climate Change (IPCC). https://www.ipcc.ch/site/ass ets/uploads/2018/03/TAR-13.pdf. Accessed 29 March 2022.

Nickless, E. (2017). Delivering sustainable development goals: The need for a new international resource governance framework. *Annals of Geophysics, 60, Fast Track 7: Geoethics at the heart of all geoscience.* https://doi.org/10.4401/ag-7426

Nurmi, P. A. (2017). Green Mining—A holistic concept for sustainable and acceptable mineral production. *Annals of Geophysics, 60, Fast Track 7: Geoethics at the heart of all geoscience.* https://doi.org/10.4401/ag-7420

Oreskes, N. (2014). The scientific consensus on climate change. *Science, 306*(5702), 1686. https://doi.org/10.1126/science.1103618

Orion, N. (2019). The future challenge of Earth science education research. *Disciplinary and Interdisciplinary Science Education Research, 1*, 3. https://doi.org/10.1186/s43031-019-0003-z

Ott, K. (2014). Institutionalizing strong sustainability: A Rawlsian perspective. *Sustainability, 6*(2), 894–912. https://doi.org/10.3390/su6020894

Peppoloni, S. (2012). Ethical and cultural value of the Earth sciences. Interview with Prof. Giulio Giorello. *Annals of Geophysics, 55*(3), 343–346. https://doi.org/10.4401/ag-5755

Peppoloni, S. (2014). *Convivere con i rischi naturali* (p. 148). Il Mulino. ISBN 978-8815250780.

Peppoloni, S., Bilham, N., & Di Capua, G. (2019). Contemporary Geoethics within the Geosciences. In M. Bohle (Ed.), *Exploring geoethics—Ethical implications, societal contexts, and professional obligations of the geosciences* (pp. 25–70). Palgrave Pivot. https://doi.org/10.1007/978-3-030-12010-8_2

Peppoloni, S., & Di Capua, G. (2012). Geoethics and geological culture: Awareness, responsibility and challenges. *Annals of Geophysics, 55*(3), 335–341. https://doi.org/10.4401/ag-6099

Peppoloni, S., & Di Capua, G. (2016). Geoethics: Ethical, social, and cultural values in geosciences research, practice, and education. In G. R. Wessel & J. K. Greenberg (Eds.), *Geoscience for the public good and global development: Toward a sustainable future* (Special Paper 520, pp. 17–21). Geological Society of America. https://doi.org/10.1130/2016.2520(03)

ProGEO (2017). *Geodiversity, Geoheritage & Geoconservation—The ProGEO simple guide.* The European Association for the Conservation of the Geological Heritage. https://www.iucn.org/sites/dev/files/progeo_leaflet_en_2017.pdf. Accessed 29 March 2022.

Raab, T., & Frodeman, R. (2002). What is it like to be a geologist? A phenomenology of geology and its epistemological implications. *Philosophy & Geography, 5*(1). https://doi.org/10.1080/10903770120116840

Ripple, W. J., Wolf, C., Newsome, T. M., Barnard, P., & Moomaw, W. R. (2020). World scientists' warning of a climate emergency. *BioScience, 70*(1), 8–12. https://doi.org/10.1093/biosci/biz088

Ripple, W. J., Wolf, C., Newsome, T. M., Gregg, J. W., Lenton, T. M., Palomo, I., Eikelboom, J. A. J., Law, B. A., Huq, S., Duffy, P. B., & Rockström, J. (2021). World scientists' warning of a climate emergency 2021. *BioScience, 71*(9), 894–898. https://doi.org/10.1093/biosci/biab079

Schwalm, C. R., Glendon, S., & Duffy, P. B. (2020). RCP8.5 tracks cumulative CO_2 emissions. *PNAS, 117*(33), 19656–19657. https://doi.org/10.1073/pnas.2007117117

Seddon, G. (1996). Thinking like a geologist: The culture of geology. Mawson Lecture 1996. *Australian Journal of Earth Sciences, 43*, 487–495.

Sonter, L. J., Dade, M. C., Watson, J. E. M., & Valenta, R. K. (2020). Renewable energy production will exacerbate mining threats to biodiversity. *Nature Communications, 11*, 4174. https://doi.org/10.1038/s41467-020-17928-5

Steffen, W., Richardson, K., Rockström, J., Cornell, S. E., Fetzer, I., Bennett, E. M., Biggs, R., Carpenter, S. R., de Vries, W., de Wit, C. A., Folke, C., Gerten, D., Heinke, J., Mace, G. M., Persson, L. M., Ramanathan, V., Reyers, B., & Sörlin, S. (2015). Planetary boundaries: Guiding human development on a changing planet. *Science, 347*(6223), 1259855–1259855. https://doi.org/10.1126/science.1259855

Stewart, I. S., & Gill, J. C. (2017). Social geology: Integrating sustainability concepts into Earth sciences. *Proceedings of the Geologists' Association, 128*(2), 165–172. https://doi.org/10.1016/j.pgeola.2017.01.002

Stewart, I. S., Ickert, J., & Lacassin, R. (2017). Communicating seismic risk: The geoethical challenges of a people-centred, participatory approach. *Annals of Geophysics, 60, Fast Track 7: Geoethics at the heart of all geoscience.* https://doi.org/10.4401/ag-7593

Wachinger, G., Renn, O., Begg, C., & Kuhlicke, C. (2013). The risk perception paradox: Implications for governance and communication of natural hazards. *Risk Analysis, 33*, 1049–1065. https://doi.org/10.1111/j.1539-6924.2012.01942.x

Wackernagel, M., & Rees, W. E. (1995). *Our ecological footprint: Reducing human impact on the earth* (p. 176). New Society Publishers. ISBN 978-0865713123.

WCED (1987). *World Commission on Environment and Development: Our common future* (p. 300). Oxford University Press. ISBN 978-0-19-282080-8.

Wessel, G. R. (2016). Beyond sustainability: A restorative approach for the mineral industry. In G. R. Wessel & J. K. Greenberg (Eds.), *Geoscience for the public good and global development: Toward a sustainable future* (Special Paper 520, pp. 45–55). Geological Society of America. https://doi.org/10.1130/2016.2520(06)

Westerhold, T., Marwan, N., Drury, A. J., Liebrand, D., Agnini, C., Anagnostou, E., Barnet, J. S. K., Bohaty, S. M., De Vleeschouwer, D., Florindo, F., Frederichs, T., Hodell, D. A., Holbourn,

A. E., Kroon, D., Vittoria, L., Littler, K., Lourens, L. J., Lyle, M., Pälike, H., … Zachos, J. C. (2020). An astronomically dated record of Earth's climate and its predictability over the last 66 million years. *Science, 369*(6509), 1383–1387. https://doi.org/10.1126/science.aba6853

White, L., Jr. (1967). The historical roots of our ecologic crisis. *Science, 155*(3767), 1203–1207.

Woo, K. S. (2017). Role of IUCN WCPA Geoheritage Specialist Group for geoheritage conservation and recognition of World Heritage Sites, Global Geoparks and other protected areas. *Geophysical Research Abstracts*, vol. 19, EGU2017-1137, EGU General Assembly 2017. https://meetingorganizer.copernicus.org/EGU2017/EGU2017-1137.pdf. Accessed 29 March 2022.

Wyss, M., & Peppoloni, S. (Eds.). (2015). *Geoethics: Ethical challenges and case studies in earth sciences* (p. 450). Elsevier. ISBN 978 0127999357. https://doi.org/10.1016/C2013-0-09988-4

Chapter 8
Geoethics and Anthropogenic Global Changes

8.1 Environmental Emergencies

In the coming decades, a number of issues of global concern will become a top priority in science, public debate, and political, legislative and decision-making processes.

Environmental emergencies affect all natural systems, without exception. There are some phenomena whose seriousness is more evident, such as the retreat of glaciers and even the disappearance of some of them, and others that are less perceptible, although their dangers are no less serious, such as the degradation of soils. Some emergencies are now recognised and alarming, such as ocean plastics pollution, while others are known but their dramatic nature is still not fully understood, such as the threat of potential conflicts for water control in a world that is warming and, in many areas, desertifying. It was only recently, during the SARS-CoV-2 pandemic that we realised, despite ourselves, how much the increased danger of zoonoses—i.e., diseases transmissible from animals to humans—in human-dominated ecosystems has been underestimated (Gibb et al., 2020).

For years, geoscientists, biologists, economists, engineers and sociologists have been working on environmental issues. The political classes are now obliged to make courageous and decisive choices on the basis of a framework of information, data, and processes that are increasingly reliable and shared, produced by science and summarised in reports by national and international bodies. The global nature of current environmental emergencies means that they must be tackled from an international perspective, finding effective forms of cooperation that go beyond the 17 SDGs and are capable of giving centrality to the dignity of the human being and guaranteeing respect for all the elements that make up the Earth system. In many cases, the solutions offered by science to these problems already exist, and politics can no longer take the liberty of ignoring them. However, in order to counteract political inaction, it is necessary to raise public awareness of ecological issues and to select competent and responsible leaders who do not simplify issues in a propagandistic manner but prove to be equal to the challenges to be faced.

© The Author(s), under exclusive license
to Springer Nature Switzerland AG 2022
S. Peppoloni and G. Di Capua, *Geoethics*,
https://doi.org/10.1007/978-3-030-98044-3_8

8.2   Anthropogenic Impact: Population Growth and Economic Systems

Ensuring enough food, energy, raw materials and drinking water for all, safeguarding human health, managing the conflicting demands of the territory, providing for soil, water and air quality, protecting ecosystems, conserving biodiversity and geodiversity as much as possible, without this meaning hindering any economic or infrastructural development needs, are all closely interlinked challenges. These are made even more complex by continuing population growth, migration and urbanisation, over-exploitation of natural resources, the unbridled consumerism of post-industrial societies and massive inequalities in wealth, health, education, political rights and access to resources.

While geosciences are an indispensable tool for quantitatively defining the planet's natural limits and allowing a healthy and safe life for the human species, in balance with available resources and ecosystems, there is no doubt that population growth, social inequalities and the unequal distribution of resources are issues that only well-prepared and responsible politics can address, including through the creation of a widespread social consensus based on knowledge. Science can make proposals, identify critical issues and provide solutions, but it is politics that must take responsibility for making well-considered and far-sighted decisions.

The data and forecast models on world population growth expressed in a report of the United Nations (UN) leave no doubt (United Nations, 2019):

> The world's population continues to grow, albeit at a slower pace than at any time since 1950, owing to reduced levels of fertility. From an estimated 7.7 billion people worldwide in 2019, the medium-variant projection indicates that the global population could grow to around 8.5 billion in 2030, 9.7 billion in 2050, and 10.9 billion in 2100.

The report shows that with an increase of more than one billion people, sub-Saharan African countries could see more than half of the world's population growth between 2019 and 2050, continuing to increase until the end of the century. In contrast, populations in East, Central and South Asia, Latin America and the Caribbean, Europe and North America are projected to peak over the same period and begin to decline before the end of the century.

The same report, in addition to highlighting an unprecedented ageing of the world's population and a continuous global increase in longevity (also an effect of advances in medicine and food availability), albeit with significant disparities in survival in the various countries and regions, stresses that:

> Continued rapid population growth presents challenges for sustainable development. The 47 least developed countries are among the world's fastest growing – many are projected to double in population between 2019 and 2050 – putting pressure on already strained resources and challenging policies that aim to achieve the Sustainable Development Goals and ensure that no one is left behind. For many countries or areas, including some Small Island Developing States, the challenges to achieving sustainable development are compounded by their vulnerability to climate change, climate variability and sea-level rise.

Significantly, the UN report emphasises the link between population growth, increased anthropogenic impact on resources, increased challenges in achieving sustainable development goals, climate change, and sea level rise. Indeed, it could not be otherwise: the general narrative on global anthropogenic change includes population growth as one of the primary causes of the degradation of natural environments.

The human impact on the planet has increased dramatically since the 1950s, after the end of the World War II. More people on the planet means an increase in demand for energy, natural resources, food, land for agriculture and urbanisation, animals for breeding, and infrastructure, as well as an increase in pollution. The envisaged scenario by the report is that the population reaches a peak and then begins a slow decline or stabilises which, in any case, implies that over the next eighty years the population is set to grow by more than 3 billion people, thus increasing by 50% by the end of the century.

These figures alone would be enough to scare governments and public opinion, but it should be added that the current environmental situation is so serious that in 2019 the European Union declared that the climate emergency[1] can no longer be postponed, in order to induce member states to take decisive actions to reduce greenhouse gas emissions and combat climate change. It is not easy to imagine a future in which the world's population will have reached 10 billion of people. These individuals will need food (Gerten et al., 2020), work, energy, infrastructure networks, industry, land for agriculture and livestock, water, and residential areas, all in quantities enormously greater than today's needs. Without going into historical examinations of the economic models of current globalised society and critical analyses of Western consumerism, this remains a fact. And it is certainly not by dreaming of cities floating on the oceans or underwater, or of extraterrestrial colonies that the future may appear more acceptable, given the scenarios we probably have to prepare for (McLean, 2020; Xu et al., 2019).

Considering that science agrees on the causal connection between anthropogenic greenhouse gas emissions and global warming (Lynas et al., 2021; Ripple et al., 2020, 2021; Oreskes, 2014), as well as between anthropogenic activities and all ongoing phenomena of environmental degradation, further population increases, if not accompanied by changes and reforms in existing economic, social, and cultural paradigms, open up the possibility of alarming future scenarios for humanity. To this we should add that, in general expectations, the energy transition appears as one of the pillars to develop an ecological conversion of modern societies. But as Bonneuil and Fressoz (2013) point out, the human history of energy is not made up of transitions but rather of additions and, thus, more simply of percentage changes in the quantities of the various energy sources that make up the energy mix used. Therefore, it is perhaps an illusion to expect a sudden and complete energy transformation in the next few years, in which fossil fuels (the energy source most responsible for the increase in atmospheric greenhouse gases and therefore for anthropogenic global warming) are

---

[1] https://www.europarl.europa.eu/news/en/press-room/20191121IPR67110/the-european-parlia ment-declares-climate-emergency. Accessed 29 March 2022.

completely eliminated. However, what is likely is a change in the percentages of the various energy sources used, with an increasing proportion of renewables.

It is also true that the current mix of energy sources depends on fluctuations in the relationship between supply and demand and the consequent continuous modulation of prices. The growing world population, which will increasingly have access to a type of welfare modelled on the Western way of life, does not rule out the possibility that over time more polluting energies will become more economically advantageous than others with a lower environmental impact. The steady increase in energy in an expanding society inevitably requires multiple sources of supply to meet the growth in demand and, therefore, more polluting energies again may become more important in the energy mix needed to meet the global increase in demand, pending the development and increased use of new cleaner energy sources such as green hydrogen (IEA, 2019; Truche & Bazarkina, 2019).

In the Western world, the real seriousness of the demographic issue is not perceived. Societies are suffering from a constant reduction in births, and political debates periodically address the issue on how to sustain the birth rate to avoid population decline. But, if we look at the rest of the world and, in particular, the African continent where furthermore an important migration phenomenon has been going on for several years, we can see the complexity and danger of the problem. Ripple et al. (2020) have no doubt, along with the more than 11,000 scientists from around the world who have signed their article-appeal: 'economic and population growth are among the most important drivers of increases in CO_2 emissions from fossil fuel combustion' and, therefore, there is a need for 'bold and drastic transformations regarding economic and population policies'.

Once again, however, the inability of politics to read scientific data and statistical analyses and to understand the predictions of mathematical models on the possible evolution of phenomena places a mortgage on the possibilities of seizing the opportunity of modern times to put human societies on new development pathways. If one thinks that technology will save the world, perhaps further doubts may arise[2]: the massive physical infrastructure of the internet requires a huge amount of energy (7% of global energy consumption), the production of which generates 4% of greenhouse gas emissions, with an estimated doubling of these percentages by 2025. This means that, as things stand, the internet is not ecologically sustainable unless we change the energy sources that power it. The increase in energy consumption for increased access to its services, the spread of digital infrastructure, and the number of users in a more crowded world, if not adequately accompanied by an energy 'revolution' in renewable sources, will only further burden the sustainability of modern societies.

---

[2] https://www.wired.it/internet/web/2019/09/05/internet-energia/. Accessed 29 March 2022.

8.3 Global Warming: Relying on Geoengineering?

There are three emblematic images of the most recent years: the CO_2 increase curve over the last 150 years,[3] a polar bear drifting on a block of ice, and the face of Greta Thunberg, a Swedish environmental activist. The link between the three is immediate: the growth of greenhouse gases and, in particular, CO_2, ongoing global warming and its threat to living species, and the younger generation and the uncertainty of future environmental conditions.

In Chapter 5, we have already discussed the youth environmental movements, which are calling for warnings from science to be heeded and for political action to be taken. The curve of CO_2 variation over time is beginning to be well known to the general public, but it is certainly not a scientific novelty of recent years. We need only recall the 5 November 1965 report *Restoring the Quality of Our Environment*, written by a group of scientists for then US President Lyndon Johnson, which summarised the risks associated with increased carbon pollution (The White House, 1965). In particular, in the section on atmospheric CO_2, the mechanism of climate change was clearly highlighted: CO_2 produced by fossil fuels was referred to as the 'invisible pollutant', as it was capable of producing significant effects on climate despite its small percentage presence in the atmosphere. It was also pointed out that the increase was caused by human activities, particularly fossil fuels, which were the only source of CO_2 added to the natural system of oceans, atmosphere, geosphere and biosphere. Human emissions had increased the amount of atmospheric CO_2 by 25% compared to the nineteenth century.

The results of the predictive model of the variations of global surface temperatures, developed in 1974 by one of the authors of the 1965 report, are surprising and discouraging.[4] The correspondence between the CO_2 values predicted by the model for the period 1915–2015 and those actually observed (provided by the US National Oceanic and Atmospheric Administration—NOAA)[5] is impressive. It is clear that politicians have completely ignored the results and suggestions of scientists for fifty years (and to some extent continue to do so), but scientists also bear a great deal of responsibility for failing to demand greater attention to environmental issues in the past decades, probably believing that their task was limited to producing accurate scientific and technical analyses and conclusions, and not fully understanding the political and social significance and value their message could have for years to come.

The scientists involved in the 1965 report also warned of a number of side effects of air pollution, rising CO_2, and resulting global warming, some of which are now

---

[3] https://www.climate.gov/news-features/understanding-climate/climate-change-atmospheric-carbon-dioxide. Accessed 29 March 2022.

[4] https://www.theguardian.com/environment/climate-consensus-97-per-cent/2015/nov/05/scientists-warned-the-president-about-global-warming-50-years-ago-today?CMP=share_btn_fb&fbclid=IwaR1f6E2M97rkDFT6uub1CtpocsC2aaLYGg5MmxQix4IYy1PVKLdLEnElooa. Accessed 29 March 2022.

[5] https://www.esrl.noaa.gov/gmd/ccgg/trends/. Accessed 29 March 2022.

becoming increasingly evident as melting polar ice caps, rising mean sea levels, warming waters and increasing acidification, and increased photosynthesis in plants. Moreover, they considered the possibility of 'causing compensatory climate change'. At the time, the mechanistic view of an Earth system dominated by linear laws of cause and effect paved the way for geoengineering techniques (Keith, 2000, 2009), i.e., those deliberate large-scale interventions on natural systems to modify the environment and reverse triggered climate change,[6] with all the consequent problems of effective feasibility and the need for responsible sharing of decision-making.

In fact, the use of fossil fuels can also be seen as a deliberate climate modification in the monumental process of constructing the human niche (Meneganzin et al., 2020), the effects of which had also been predicted in 1977 by some US oil companies but which reacted at the time by defending their interests and profits with an ad hoc disinformation campaign denying the anthropogenic causes of global warming linked to the use of oil products. The strategy adopted was aimed at inducing doubts about scientific findings to confuse the public, instilling in society the view that scientists have no certainty, and promoting environmental initiatives as diversions to enhance their social reputation (Cook et al., 2019; Oreskes & Conway, 2010).

This episode clearly shows the resistance of some sectors of society, aimed at maintaining the economic and energy status quo in spite of the warnings from science. And now that global warming is beginning to show itself in a tangible way, the urgency of finding solutions of a different kind is also reviving the debate on the possible use of geoengineering. This is despite the fact that its use in the military sector was already banned in 1978 by a United Nations convention,[7] after the United States used cloud seeding during the Vietnam War to artificially cause rainfall and thus make the communication routes used by the Viet Cong in the jungle impassable (Bonneuil & Fressoz, 2013).

Geoengineering techniques can be divided into two groups: those related to solar geoengineering, which aim to manage solar radiation by increasing the earth's albedo, i.e., the earth's reflectivity towards sunlight, and those used in carbon geoengineering, which aim to remove greenhouse gases from the atmosphere. Some of the best-known geoengineering techniques include the following: reforestation/afforestation, i.e., the planting of trees on a global scale to fix atmospheric CO_2; the use of biochar, a charcoal produced by pyrolysis of plant biomass, which can trap more CO_2 than it emits for energy production; to reduce emissions of the greenhouse gas nitrogen dioxide (NO_2) from the soil and to improve the physical and chemical characteristics of soils[8]; the cultivation of biomass to capture atmospheric CO_2, which is then burned to develop energy by sequestering the CO_2 emitted, thus preventing it from being reintroduced into the atmosphere; the construction of plants to capture CO_2

---

[6] http://www.geoengineering.ox.ac.uk/what-is-geoengineering/what-is-geoengineering/. Accessed 29 March 2022.

[7] Convention on the prohibition of military or any other hostile use of environmental modification technique, New York, 10 December 1976. The Convention entered into force on 5 October 1978: https://treaties.un.org/Pages/ViewDetails.aspx?src=IND&mtdsg_no=XXVI-1&chapter=26&clang=_en. Accessed 29 March 2022.

[8] https://regenerationinternational.org/2018/05/16/what-is-biochar/. Accessed 29 March 2022.

directly from the atmosphere, then store it in underground deposits (Carbon Capture and Storage—CCS); oceanic fertilisation, i.e., the addition of nutrients to water in selected locations to increase its ability to trap CO_2; increasing meteoric degradation, by exposing large quantities of minerals to the air, which react chemically with atmospheric CO_2 and can then be stored in the soil or in the oceans; and, finally, improving ocean alkalinity by pulverising, dispersing, and dissolving rocks such as limestone, silicates or calcium hydroxides in the oceans to increase their carbon storage capacity and decrease their acidification.

From the point of view of geoethics, employing geoengineering techniques instead of focusing on activities to reduce the emission of pollutants into the atmosphere only diminishes human responsibility towards the planet, replacing it with a hoped-for saving action of technology and thus increasing the moral hazard of society (Dembe & Boden, 2000; Morini, 2014). The 'saving' vision of technology might push humans to continue polluting or even increase pollution on the planet, as ultimately they will consider geoengineering capable of solving all the environmental problems and will feel relieved of any responsibility for their impacting actions and behaviours.

Several other ethical issues arise in the use of geoengineering. Firstly, solar geoengineering interventions, even if implemented by a single nation, would have transnational effects, as well as producing benefits for some countries or regions but disadvantages for others. The need to establish decision-making chains with shared responsibility and to ensure fairness in the distribution of pros and cons in the application of these technologies would require agreements between governments and policies of distributive justice and social protection of the poorest who, although responsible to a lesser extent for the human impact on the planet (Kartha et al., 2020), might also have to bear the negative consequences of the interventions adopted. However, the most worrying aspect of the use of such techniques derives from the fact that the deliberate alteration of a natural system always entails partly unforeseeable and inevitable consequences on environmental balances, consequences that would be borne by future generations. The possibility that these actions mortgage the future of Earth's next inhabitants calls for a high sense of intergenerational responsibility. Geoengineering is a fundamentally ethical and political issue, even before being technical (Burns, 2011; Preston, 2013), a field of heated debate between different visions of the world to come. Humanity runs the risk of not having enough time to decide what to choose for the future, and that our procrastination will force us to suddenly resort to technology for having wasted too much time.

In the end, Kiehl (2006) is probably right when, in questioning the proposals of chemist and Nobel Prize Paul Crutzen to intervene with solar geoengineering to stop global warming (Crutzen, 2006), while recognizing the need for a debate of scientific ideas without prejudice, says: 'On the issue of ethics, I feel we would be taking on the ultimate state of hubris to believe we can control Earth'; and adds: 'I feel that treating the cause(s) rather than the symptom is the more appropriate approach to the problem'. Geoengineering would work on the symptoms, while more responsible human beings, each for their own possibilities, would affect the causes. Could the aim be enough to lean towards these technologies?

8.4 Natural Hazards: Are They Increasing?

Earthquakes, landslides, volcanic eruptions, floods, cyclones and rainfall are evidence of a living planet, subject to incessant matter and energy flows that interact with human beings and their activities. According to Swiss Re, a Swiss insurance company, in 2017 there were more than 11,000 victims of natural and man-made disasters, with economic losses estimated at least at USD 306 billion,[9] while in 2018 there were 13,500 victims and economic losses of USD 165 billion. Since 1970, the natural events that have produced the largest number of victims in the world have been Cyclones Bhola (1970) and Gorky (1991) in Bangladesh, the Tangshan earthquake in China (1976), the Indian Ocean earthquake and subsequent tsunami (2004), Cyclone Nargis in Myanmar (2008), the Haiti earthquake (2010), Typhoon Haiyan in the Philippines (2013), and the Nepal earthquake (2015).[10] In 2012, experts from Munich Re, a German insurance group, pointed out that in the near future environmental and climatic disasters, particularly floods and droughts, will have the greatest impact on human communities (Peppoloni, 2014). These considerations were also confirmed by the World Economic Forum in its Global Risks Report (WEF, 2020).

On Earth there are about 550 active volcanoes (that is, they have produced eruptions in historical times) and, among them, more than 50 erupt every year and another 1,000 are potentially active.[11] In addition, there are 'supervolcanoes' such as the Campi Flegrei or Yellowstone, which are capable of generating gigantic eruptions and ejecting so much solid, fluid, and gaseous material into the atmosphere that they cause planetary climatic changes, as well as death and destruction for tens of kilometres around the eruption zone. For the period 1980–2018,[12] Munich Re has estimated total damages from volcanic eruptions to be USD 12 billion. As for damages caused by hurricanes, the figures are USD 267 million in 2005 and 230 million in 2017, the year in which Hurricanes Harvey, Irma, and Maria caused record economic losses in just four weeks. Similar estimates are available for other risks triggered by natural phenomena, such as wildfires, storms, tornadoes, floods, droughts and heat waves.

From an economic perspective, in its 2018 annual report, Munich Re considers climate change to be among the biggest long-term risks for the insurance industry, as it will lead to an increase in extreme weather events and, consequently, associated disasters (Munich Re, 2019). In fact, overall economic losses caused by disasters are increasing worldwide[13] and it is clear that climate change will heavily affect

---

[9] https://www.swissre.com/dam/jcr:e41e80c0-e651-4efd-bf17-a5e1ec5109ef/nr20171220_s igma_estimates.pdf. Accessed 29 March 2022.

[10] https://www.swissre.com/dam/jcr:c37eb0e4-c0b9-4a9f-9954-3d0bb4339bfd/sigma2_2019_en. pdf. Accessed 29 March 2022.

[11] https://www.usgs.gov/faqs/how-many-active-volcanoes-are-there-earth?qt-news_science_pro ducts=0#qt-news_science_products. Accessed 29 March 2022.

[12] https://www.munichre.com/en/risks/natural-disasters-losses-are-trending-upwards/volcanic-eru ptions-the-earths-ring-of-fire.html. Accessed 29 March 2022.

[13] https://www.munichre.com/topics-online/en/climate-change-and-natural-disasters/natural-dis asters.html. Accessed 29 March 2022.

economic dynamics (Auffhammer, 2018) and the stability of the global banking system (Lamperti et al., 2019).

The common perception is that natural phenomena are increasing in frequency and intensity. However, in the case of earthquakes[14] and volcanic eruptions,[15] the occurrence rates are unchanged, as they depend on geodynamic characteristics of the Earth system that change only over hundreds of thousands/millions of years. On the other hand, in the case of phenomena such as heat waves, tornadoes, very intense and concentrated rainfall, and large hurricanes, the perception seems not to be wrong. The greater amount of energy present in the outer shell of the planet due to global warming, the changed thermal, water, and rainfall characteristics, the increase in thermal contrasts between different continental areas and between these and the oceans are producing, in some cases, an increase in the number of events or an increase in so-called extreme events, characterised by strong intensity. In addition to these phenomena, there are also droughts and large-scale fires, such as those that occurred in 2019 and 2020 in Siberia, Canada, California and Australia. Looking at the number of disasters, which Kelman (2019) highlights to be essentially social events triggered by natural phenomena, with a few exceptions there has been an increase since 1970, with an acceleration since 1980.[16] Natural phenomena are thus having an increasing societal impact.

If we look at the poorest countries on the planet, we can see how close the relationship is between natural hazards and social inequalities. According to the World Bank, around 1 in 10 people worldwide live on less than USD 2 a day (internationally recognized threshold for extreme poverty) and 85% of these poor people are concentrated in the top 20 countries most vulnerable to climate change.[17] In low-income countries, widespread urbanisation tends to concentrate the population in very vulnerable settlements, where the urban and construction fabric does not adequately take into account the hazards of the territory. This significantly increases the risk for the population, both for the direct consequences caused by the occurrence of a natural event and for the indirect consequences due to the difficulties in bringing help to those affected after the event.

In general, the increase in population in a place leads to an increase in the risk associated with natural phenomena. There is an increase in exposure, i.e., the number of people who may be affected by the event and the services that must guarantee their assistance, sustenance, and mobility, and new production and settlement areas progressively occupy more dangerous areas from a geological and climatic point of view, which had been carefully avoided in the past. It should be added that it is the globalized society as a whole that has become more vulnerable. Supply chains, infrastructure networks, and close economic relationships between states have created a

---

[14] http://www.earthquakes.bgs.ac.uk/news/EQ_increase.html. Accessed 29 March 2022.

[15] https://volcano.si.edu/faq/index.cfm?question=historicalactivity. Accessed 29 March 2022.

[16] https://www.swissre.com/dam/jcr:c37eb0e4-c0b9-4a9f-9954-3d0bb4339bfd/sigma2_2019_en.pdf. Accessed 29 March 2022.

[17] https://www.weforum.org/agenda/2019/09/how-survive-to-a-natural-disaster-with-less-than-2-per-day/. Accessed 29 March 2022.

seamless planetary system that can more easily mobilise and react to absorb the effects of a disaster but can also suffer its negative repercussions in a faster and more widespread way.

In this context, it is clear that an effective risk reduction policy must be based on a comprehensive prevention strategy, capable of addressing all the elements that contribute to the definition of risk itself, from the local to the transnational scale, including engineering, geological, and urban-architectural planning aspects, as well as sociological, medical, economic and political elements. A short article published on the World Economic Forum website[18] reports that, on average, for every Euro invested in activities that increase the resilience of poor countries, between four and seven Euros can be saved in subsequent reconstruction costs, not to mention that investments in risk reduction generate a multiplier effect of attracting further capital for development. However, the same article indicates that, according to the United Nations, only 3.8% of official development programmes directly support disaster preparedness, when instead the only way to reduce the impact of disasters is to get ready before they happen in order to reduce the vulnerability and exposure of affected communities. Ultimately, not investing in prevention, especially for high-income countries that would have available resources, means irresponsibly shifting the social and economic cost of disasters onto the shoulders of the poorest people and future generations (Peppoloni, 2014).

8.5 Declining Biodiversity: Are We Close to a New Mass Extinction?

A number of authors who lived between the nineteenth and twentieth centuries assumed, each from his own particular perspective, that our planet is an organism with its own physiology and metabolism and acts as a living and 'thinking' being.

Pierre Teilhard de Chardin (1881–1955), a French philosopher and palaeontologist, with the formulation of the *Law of Complexity-Consciousness*, and in agreement with the Russian-Ukrainian geochemist Vladimir Ivanovič Vernadskij (1863–1945), theorised on the tendency of matter to become increasingly complex as it moves from the geosphere (inanimate matter) to the biosphere (living matter) to the noosphere (conscious matter). He considers the noosphere to be a kind of collective consciousness of human beings, the end point of the historical evolution of nature and consciousness, a concept that once again recalls Élisée Reclus' statement at the beginning of the twentieth century (see Chapter 2).

In 1979, the British chemist James Lovelock developed the 'Gaia hypothesis', in which Earth is conceived as a complex, living, self-regulating superorganism, subject to evolutionary phenomena that allow the maintenance of life and that can be studied from the perspective of a new transdisciplinary science—geophysiology (Lovelock,

---

[18] https://www.weforum.org/agenda/2019/09/how-survive-to-a-natural-disaster-with-less-than-2-per-day/. Accessed 29 March 2022.

1990). But his vision seems to re-propose the fracture between humanity and nature, since human beings are considered outside the system and their activities a polluting factor for the superorganism.

Over time, the hypothesis of Earth as a single living system has become the conceptual paradigm of reference for many environmentalist visions. The Russian biochemist and writer Isaac Asimov (1920–1992), in his 1986 science fiction novel *Foundation and Earth,* conceives of Earth as a conscious planet, 'Gaia', in contact with all its life forms, including humans, and imagines each living being in turn in relation to every other being and to the planet itself. In the novel, this holistic vision is contrasted with psychohistory, an imaginary mathematical-statistical science that predicts social but not individual behaviour on a probabilistic basis, provided certain boundary conditions are met.

Lovelock's 'Gaia hypothesis', which was initially criticised for its cooperative view of organisms in contrast to the evolutionist theory of competition between living species, is countered by the 'Medea hypothesis' of the paleontologist Peter Ward. He sees the extinctions that have occurred repeatedly on Earth as a natural 'biocidal' tendency of life (Ward, 2009) and also frames the ongoing anthropogenic change in this hypothesis. The destructive action of humans towards ecosystems and, ultimately, towards themselves would, therefore, be nothing new in the geological history of the planet.

It is obvious that respecting the Earth system means, first and foremost, respecting life, bearing in mind that in any case the earth will be able to reabsorb the consequences of irresponsible human activities and that what is really at risk on the planet is the quality of human existence itself, if not even its survival and that of many other living species. Lovelock himself speculates that the current world may have passed the threshold beyond which it is no longer possible to prevent changes caused by human activity from destroying civilisation (Lovelock, 2006). The anthropogenic loss of biodiversity, which in Ceballos et al. (2017) is defined as 'biological annihilation', would be evidence of an ongoing global ecosystem collapse ('the sixth great mass extinction on Earth' for Ceballos et al., 2015; actually it is the seventh, since recent studies, in Dal Corso et al., 2020, have identified another mass extinction in the Upper Triassic), which would also include human living space. The extreme consequence of such a process would be the inevitable destruction of human civilisation, which quite unreasonably still believes that continuous and uninterrupted population growth and the extractive activities necessary to sustain it are still possible on a planet that is instead limited in its resources.

Biodiversity is the variety of living entities that inhabit Earth (plants, animals, and microorganisms) and is measured at the level of genes, species, populations and ecosystems.[19] Together with geodiversity, it contributes to providing us with the ecosystem goods and services that nature makes available to life on the planet. These are fundamental to our well-being, health, and the economic wealth of society, contributing to the creation of natural capital—food, drinking water, timber, fibers,

---

[19] https://wwfit.awsassets.panda.org/downloads/lpr_2018_ita_highlights_1.pdf. Accessed 29 March 2022.

fuel, plants but also mechanisms for the management, treatment and purification of nutrients and waste, regulation of infectious diseases and climate, or the recreational, cultural, and spiritual services that are the intangible benefits of ecosystems for human beings. In these terms, the loss of biodiversity leads to degradation of ecosystem services, with a consequent increase in food, health, water, energy, natural and socio-cultural risks to society (Millennium Ecosystem Assessment, 2005).

In the report of the World Wildlife Fund (WWF) entitled *Living Planet Report 2018*, it is stated that 'Despite well-meaning attempts to stop this loss through global agreements such as the Convention on Biological Diversity'[20] the global trend of biodiversity destruction continues unabated, making pursuit of the 17 SDGs impossible. Even if we wanted to completely neglect the emotional and psychological aspect that binds humans to nature and the life forms that make it up, to remain solely within the economic logic of monetisation underlying the concept of natural capital, the WWF report estimates that 'globally, nature provides services worth around USD 125 trillion a year', a figure higher than the gross global product of countries around the world, which is around USD 80 trillion. It adds, 'Governments, business and the finance sector are starting to question how global environmental risks – such as increasing pressure on agricultural land, soil degradation, water stress and extreme weather events – will affect the macroeconomic performance of countries, sectors and financial markets'. But the WWF report is enlightening and, in many ways, disturbing: agriculture, which is necessary to sustain a growing world population, is the main threat of extinction for 8,500 of the living species included in the Red List of the International Union for Conservation of Nature (IUCN), and 75% of the extinctions of vertebrate plants and animals that occurred since 1500 AD can be attributed to agricultural activities. Hence, the urgent need to adopt more sustainable agricultural practices.[21]

In addition, pollution and resulting climate change, dams, mining, forced relocation of living species for commercial reasons and along global supply chains in general—a phenomenon that accelerated dramatically at the end of the fifteenth century, after the arrival of Christopher Columbus on the American continent (Lewis & Maslin, 2018) and has had a further increase due to globalisation in the last thirty years—are factors responsible for an increase in the ecological footprint of 190% in fifty years and, therefore, for the continuous loss of biodiversity at a planetary level, albeit with significant differences between high- and low-income countries. Furthermore, the loss of biodiversity affects all natural environments, including soil, which alone is home to around 25% of all life on Earth and contributes to the absorption of atmospheric CO_2 by acting on nutrient cycles. Life on the planet is so interconnected that it is up to the human being to start thinking holistically because, in fact, nature is already holistic and has always been so.

---

[20] https://www.cbd.int/convention/text/. Accessed 29 March 2022.

[21] https://www.greenmatters.com/t/sustainable-agriculture. Accessed 29 March 2022.

8.6 Soil Degradation and Water Pollution: Causes of Future Conflicts?

The 2018 Assessment Report on Land Degradation and Restoration, produced by the Intergovernmental Science-Policy Platform on Biodiversity and Ecosystem Services (IPBES), highlights the urgency of combating soil degradation to protect biodiversity and ecosystem services and, thus, the well-being of human life (IPBES, 2018). Soil degradation is defined as 'the loss of land's production capacity in terms of loss of soil fertility, soil biodiversity, and degradation. Soil degradation causes include agricultural, industrial, and commercial pollution; loss of arable land due to urban expansion, overgrazing, and unsustainable agricultural practices; and long-term climatic changes' (Maximillian et al., 2019). It affects all natural environments on the planet, all regions and biomes, although the severity of the phenomenon varies. There is no system modified by humans (arid zone, agricultural and forestry system, savannah or aquatic system) that is not affected by this phenomenon. Soil degradation has direct effects on 3.2 billion people, with an estimated economic loss of 10% of global gross annual product.

Therefore, taking action to reduce soil degradation has a crucial environmental, social, and economic significance for the pursuit of the 17 SDGs while, at the same time, having positive impacts in terms of increasing the quantity and quality of food and water, increasing employment, reducing social inequalities, conflicts and migrations that mainly affect the poorest parts of the planet. By now, less than 25% of Earth's surface is in natural condition and it is estimated that, at the current rate of Earth's overexploitation, it will drop to 10% by 2050. Within sixty years, all of the world's topsoil, rich in organic matter and microorganisms, composed of mineral particles, air, and water, could become unproductive (Maximillian et al., 2019).

Land degradation also includes the loss of forests, a phenomenon that, albeit contrasted in temperate areas by reforestation projects, has accelerated in tropical areas, where the development of both subsistence and intensive agriculture led to the conversion of 40% of forests to agricultural land between 2000 and 2010. Another 27% of the loss is due to urban growth, infrastructure development, and mining. Loss of biodiversity and soil erosion, pollution and nutrient surplus, intensive agriculture and livestock farming, fires, desertification and climate change are therefore considered the main causes of soil degradation.

In the absence of global initiatives, this phenomenon is likely to intensify further as a result of the growth of the world's population and resource consumption in both developed and developing countries, and the emergence of an increasingly globalised economy, which facilitates access to world markets for all local players while reducing the perceived impact of one's own consumption due to the greater distance between areas of production and consumption.

Technological innovation, sustainable agriculture, and a reduction in intensive livestock farming are the answer to the problem which, in any case, requires international agreements to regulate production and supplies, to reduce pressure on ecosystems and allow a gradual recovery of their physical-chemical-biological characteristics. From this point of view, the introduction of hydroponic crops (even if, at the moment, it is only feasible at high cost and with the use of a proportion of non-recyclable materials) is an interesting evolution of agricultural techniques, as crops are grown above ground or without soil, providing the replacement of soil with various inert materials (such as expanded clay and zeolites, i.e., sodium, calcium and potassium silicates), a high degree of automation and control of the energy and water used (saving up to 90%) and the use of nutrients from low-cost recycled organic material.

Therefore, sustainable solutions are already feasible with adequate investment and precise political choices of development for the future. Increasing attention is paid to sustainable agriculture, which the Food and Agriculture Organization (FAO) of the United Nations promotes, based on five principles[22]: (1) increasing productivity, employment, and value addition in food systems; (2) protecting and enhancing natural resources; (3) improving livelihoods and fostering inclusive economic growth; (4) enhancing the resilience of people, communities, and ecosystems; and (5) adapting governance to new challenges. The adoption of these principles aims not only to safeguard the productive layer and, more generally, the environment, and reduce soil erosion (thereby helping to stabilise slopes otherwise affected by landslides or to counteract surface water runoff by increasing infiltration and recharging of groundwater aquifers) but also to ensure the social sustainability of farming communities and the profitability of activities over time.

Unfortunately, the current reality is that the soil is becoming progressively impoverished and, once it has become unproductive, it will not allow the growth of plants, i.e., those organisms that, in addition to providing sustenance for living species, are capable of fixing atmospheric CO_2 through photosynthesis. In order to sustain soil productivity, industrial fertilisers are used on a massive scale, causing the growth of harmful algae which, in turn, produce neurotoxins and destroy plants, creeping into the food chains of animals and, finally, returning to humans in a fatal cycle. Once again, locally produced human impact has extreme consequences on much larger portions of the natural system initially affected. When the effects of anthropogenic actions reach areas far removed from the place where they were carried out, the problem can widen to the point of generating conflicts between different communities.

These considerations become particularly evident when considering water, which by its nature is not spatially circumscribable by national boundaries but can cross different states and, therefore, obliges transnational management (Abrunhosa, 2021; Chaminé et al., 2021; Groenfeldt, 2020). The *World Water Development Report 2019* of the United Nations devotes a paragraph to transboundary water resources and water-related conflicts (WWAP, 2019). There are 286 international rivers and

---

[22] http://www.fao.org/sustainability/background/en/. Accessed 29 March 2022.

592 transboundary groundwater aquifers in the world, shared by 153 nations. In the period 2010–2018, at least 263 conflicts were recorded. For 123 of them, water was a direct cause of contention; in 29 cases, water was used as a weapon for political pressure, while in 133 this resource was damaged by the conflict. Although there is no substantial increase in the number of 'water wars', the report does not exclude that in the more or less near future, when the effects of climate change begin to intensify, uncertainty in the supply of resources, combined with political and social instability, could be the trigger for an increasing number of conflicts (Borgomeo, 2020).

Another relevant fact mentioned in the UN report is that the lack of drinking water and adequate sanitation still claims many more victims than floods, droughts, earthquakes, epidemics and wars. The management of water resources, access to drinking water and sanitation are issues that, in addition to technical and scientific aspects, involve questions of social equity and intergenerational justice. For example, groundwater is a renewable water resource provided that proper management based on scientific knowledge ensures that withdrawals do not exceed intakes. For this reason, it is necessary to define the extent of aquifers, their storage capacity, recharge areas due to rainfall, and the number and extent of springs that will be used for abstraction. The system must be constantly monitored and the recharge areas safeguarded in order to avoid pollution of the water entering it. Similarly, excessive water extraction from wells, which cannot be compensated for by rainwater infiltration and which, over time, leads to almost irreversible phenomena of soil compaction, must be avoided. In addition, uncontrolled water abstraction in coastal areas can lead to the mixing of fresh water with salty sea water and the irreversible salinisation of the aquifers, which can then no longer be used for drinking or irrigation purposes.

In the case of rivers, excessive direct withdrawal of water from the riverbed or splitting up into smaller channels that take water away from the main course inevitably lead to a radical and irreversible change in the natural systems in which the river is inserted. In this respect, the case of the Aral Sea, located on the border between Uzbekistan and Kazakhstan, is exemplifying. Since the 1960s, in the middle of the Soviet era, the development of intensive agriculture, together with a misguided management of the lake's tributaries, led to its gradual drying up (Izhitskiy et al., 2016). Today, the Aral Sea has shrunk to less than a quarter of its original size, its salinity has increased significantly, the surrounding ecosystems have collapsed, and the local economy has been completely destroyed. There are plans to restore the lake to its original size, but the lost ecosystems will probably not be able to be recreated. New ecosystems would be born in response to yet another change in local physical, chemical, and biological conditions. The fact remains that the Aral Sea area has been subjected to some of the greatest anthropogenic environmental destruction, and perhaps leaving it as it is, as a warning for the future, might be more effective and much less costly than trying to restore it to its original condition.

Today, the world's population consumes 4,600 cubic kilometres of water per year, of which 70% is used for agriculture, 20% for industry and 10% for domestic use. Water is an increasingly endangered resource. The above-mentioned UN report predicts that:

Global water demand is expected to continue increasing at a similar rate until 2050, accounting for an increase of 20 to 30% above the current level of water use, mainly due to rising demand in the industrial and domestic sectors. Over 2 billion people live in countries experiencing high water stress, and about 4 billion people experience severe water scarcity during at least one month of the year. Stress levels will continue to increase as demand for water grows and the effects of climate change intensify.

The same report indicates that:

Three out of ten people do not have access to safe drinking water. Almost half of people drinking water from unprotected sources live in Sub-Saharan Africa. Six out of ten people do not have access to safely managed sanitation services, and one out of nine practice open defecation. However, these global figures mask significant inequalities between and within regions, countries, communities and even neighbourhoods.

Water remains an inalienable human right, a guarantee of the dignity of every individual. Although each community or state has the legitimate right to develop policies that safeguard its own interests and priorities, no one can contravene that fundamental right, which belongs to the human species and to all forms of life on Earth that depend on that resource.

8.7   Mineral Resources: Circular Economy or Extraction from the Oceans?

It is estimated that there are more than 5 billion mobile phone and computer users in the world.[23] This number gives the sense of the penetration of digital technology into our daily lives. The reference object par excellence of this phenomenon is the smartphone. Not many people are aware that to make a smartphone, numerous minerals are needed. Those minerals are obtained through an impressive industrial, technological, and commercial chain that consumes large amounts of energy and produces pollution. The activity of extracting the necessary minerals has a major ecological impact, as does the construction of the infrastructure network for transporting the raw materials, the industries for processing the minerals into components to be assembled, the networks for transporting and distributing the finished products, and the facilities for disposing of disused devices.

Rare earths, a group of 17 chemical elements from the periodic table that are extracted from certain rocks, including carbonatites (igneous rocks) and lateritic clays (residual rocks), are used to make the color screens, the electrical circuits, the microphones, the vibration mechanism of smartphones. Despite their name, rare earths are metals (electrical and heat conductors) that are quite widespread in the earth's crust, although they are difficult to extract and are sometimes present in deposits with concentrations too low for their economically viable exploitation. Rare earths are also used to make superconductors, magnets, optical fibres, chemical catalysts,

---

[23] https://wearesocial.com/uk/blog/2019/01/digital-in-2019-global-internet-use-accelerates/. Accessed 29 March 2022.

metal alloys, car alternators and computers. About 80% of them come from geological deposits in China, while other significant deposits are found in Brazil, Australia, Canada and the United States. The production of a smartphone also requires about 9 g of copper, 250 mg of silver, 24 mg of gold and 9 mg of platinum. In addition, a lithium-ion battery that powers an electronic device contains about 3.5 g of cobalt and 1 g of rare earths. These few figures give a partial idea of how great the impact of human activity is on the earth's crust. Obtaining the elements needed to build a single smartphone requires significant quantities of raw materials, which currently come almost exclusively from mining.

A first consideration is that the demand for rare earths has increased significantly since the 1950s, but their presence in economically exploitable mineral deposits remains limited. Moreover, rare earth mining is an environmentally harmful and toxic activity, since it is accompanied by the production of tens of millions of tonnes of waste water and the mobilisation of millions of tonnes of rocks from which to extract the necessary metals. If we consider that in the coming years the demand for rare earths will continue to increase due to the growing demands of the electric car industry, energy-efficient lighting, and medical diagnostics (not to mentioning the demands for use in new land and marine armaments), we can easily understand the ethical dilemma that we, as a society, will have to face in order to find a balance between the development of these new technologies and the conservation of the ecosystems from which the necessary natural resources come.

But, the complexity of the existing problems is such that even defining realistic scenarios is quite difficult, due to the constant change of the variables at stake. By way of example, while the high-tech industry and the industry of clean energy production equipment (such as wind turbines) has driven the growth of the rare earths market, in 2010–2011 China introduced export restrictions on these metals to preserve available resources (Mancheri et al., 2019), which has led to a rise in price. In addition, the rate of growth in global demand has not been offset by opening an adequate number of new sites for mining, and the resulting reliance on reserves has reduced their capacity.

The situation changed in 2013, when China lifted export restrictions and increased production. There are more than 850 rare earth deposits in the world, and maintaining the current rate of production could meet demand for over a hundred years, albeit with marked differences between different rare earths, considering that in some cases demand for individual elements will fall, while for others it will rise (Zhou et al., 2017).

From a geopolitical point of view, there is an objective problem of inhomogeneity in the geographical distribution of deposits on the planet, a distribution linked to local geo-lithological characteristics. This means that many countries find themselves in a position of strong dependence on a few producing countries. Without the common will to establish international governance of the strategic rare earths market, capable of preventing tensions and conflicts between states, import-dependent nations will be forced to polarise around the few producing nations in a pattern of large blocs in strong political, economic, commercial and military competition.

Such a complex and interdependent economic system—in which the supply chains of raw materials and goods that are now indispensable for the development and maintenance of current economic and social structures can be conditioned by individual international actors, such as China, which holds the record for world production of rare earths (Mancheri et al., 2019)—calls for a sense of responsibility in world politics that has never been experienced in the past. It also calls for the development of technologies to recover raw materials from unused processing products, which will progressively reduce reliance on exploration and exploitation of new mining sites and dependence on producing countries. Unfortunately, at present, only limited quantities of rare earths can be recovered from batteries, permanent magnets and fluorescent lamps and, in any case, recycling costs remain high and uneconomic.

While this is true for rare earths, it is not the case for many other minerals and metals used by industry to produce goods and services. Iron, copper, and aluminium are the most widely used metals in the world, accounting for around 95% of annual industrial metal production, and demand for them is expected to triple by 2050. The increasing economic, geopolitical, environmental and social risks of mining investments pose serious problems of sustainability of supply systems for these metals to meet future market demands, unless major changes occur in the industrial and more generally economic system (Lèbre et al., 2019).

With increasing frequency one hears about the circular economy, an expression that the private American foundation Ellen MacArthur[24] uses to redefine growth, focusing on positive benefits for the whole of society. It 'entails gradually decoupling economic activity from the consumption of finite resources, and designing waste out of the system. Underpinned by a transition to renewable energy sources, the circular model builds economic, natural, and social capital'. In the circular economy, reference is made to the non-linear feedback mechanisms of biological systems, whereby the raw materials used are subsequently fed back into both the biological and technical cycles.

The circular economy is regenerative, based on renewable energies, and stands in contrast to the current linear economic model, which is based on take-make-dispose concepts and is fueled by fossil fuels. In the linear economy, at the end of the supply-production-consumption chain, pollutants and waste which are very often non-biodegradable, such as oil-derived plastics, are released into the Earth system. In the circular economy, the economic system must be structured to design and produce objects that can be completely recycled at the end of their life cycle. This means that the economic system will no longer need to take natural materials from the Earth system to maintain a certain level of production because the economy of products will be based exclusively on recycling what has already been produced.

A circular economy is able to reduce CO_2 and energy emissions, as there is no need to transform raw materials into engineered elements. It also promotes the regeneration of ecological systems, as the impact on these is limited to the introduction into the economic system only of the raw materials needed to satisfy an initial growth in

---

[24] https://www.ellenmacarthurfoundation.org/circular-economy/concept. Accessed 29 March 2022.

demand for specific goods and services. It results in a kind of controlled design of the small percentage of waste and pollution produced in the life cycle of objects. Finally, in a perfectly circular economy, there is no waste and no pollution.

Unfortunately, the current economic paradigms are a long way from circularity. The amount of waste produced is enormous and the packaging of products is not designed to meet the criteria of a circular economy. Waste can only be partially recycled through separate collection which, in order to work, must be widespread, efficient and on time. Entire sectors of the economy are still based on uncontrolled consumerism, in which planned obsolescence, a production strategy that has been denounced for decades, is actually used in some sectors. That this is not always one of many conspiracy theories seems to be confirmed by a number of class actions in the United States in 2003, which drew general attention to this business model, which has been known since 1924 and to which the European Union has also devoted particular attention with a view to its definitive ban.[25]

The recycling-driven economy is the great opportunity for the future and, in this respect, mining itself will be at a crossroads in the coming years. The percentage of some metals that can be recovered from discarded smartphones (in this case we can speak of 'urban mining') is much higher than the amount that can be extracted from an equivalent weight of rocks. About one gram of gold can be extracted from one tonne of rocks, while 25 g can be extracted from one tonne of smartphones. The European Union is investing significant financial resources in the recovery of essential and other raw materials from landfills and mining waste.[26] In addition, robotization will make it possible to reopen mining in flooded mines[27] (Chakravorty, 2019) or to deepen mining sites that were abandoned in the past due to the inability to guarantee acceptable safety conditions for the staff working there. Finally, the refinement of geophysical prospecting techniques and methods of analysing data already collected in the past will make it possible in the near future to reopen closed mining sites or to breathe new life into mines in the process of being depleted, thus significantly limiting the impact of opening up new areas for mining and facilitating social acceptance of these activities.[28]

However, it must be made clear that mining is the basis for the creation of any human artefact and is therefore essential for the maintenance of civilisation. But, while there is a growing demand for more sustainable practices that respect the environment and local communities, and this is leading to the concept of responsible mining[29] (Arvanitidis et al., 2017; Bilham, 2020; Boon, 2020a, 2020b; Mudd, 2020), current economic logic is pushing and widening the boundaries of exploration and

---

[25] https://www.europarl.europa.eu/doceo/document/E-8-2018-001864_EN.html. Accessed 29 March 2022.

[26] https://op.europa.eu/en/publication-detail/-/publication/4b410d88-a774-11e9-9d01-01aa75ed7 1a1/language-en. Accessed 29 March 2022.

[27] https://www.unexmin.eu/. Accessed 29 March 2022.

[28] https://smartexploration.eu/. Accessed 29 March 2022.

[29] https://responsibleminingindex.org/en; https://miningwithprinciples.com/. Accessed 29 March 2022.

exploitation. This is the case of so-called deep sea/ocean mining. Mineral deposits in the oceans have the potential to provide society with many of the raw materials needed to meet a growing global demand (e.g., cobalt, manganese, thallium, nickel, gold). The oceans occupy 70% of Earth's surface and this figure alone is enough to understand the economic potential of seabed mining.

As early as the nineteenth century, the discovery of polymetallic nodules on the seabed hinted at possible applications for future exploitation. From the 1950s onwards, researchers began to identify mineral-rich oceanic hydrothermal springs (known as 'black smokers'), particularly metal sulphides, metal slurries and poly-metallic encrustations. Between the 1960s and 1980s, a number of mining companies began planning to harvest nodules in the Pacific Ocean from which to extract the minerals of interest, but a worldwide decline in commodity prices and uncertainty over the ownership rights of submarine deposits in international waters halted these projects. In the 1990s and early the twenty-first century, again in the Pacific Ocean (in Papua New Guinea), the discovery of vast new areas of hydrothermal springs rich in metals and minerals rekindled interest in their commercial exploitation. In addition, since 1994, an international body, the International Seabed Authority (ISA),[30] has been active, founded following the entry into force of the Convention on the Law of the Sea[31] of the United Nations, to coordinate and control all activities related to minerals on the international seabed (which covers more than 54% of the total area occupied by the oceans), i.e., beyond the boundaries of national jurisdictions.

In the case of deep sea/ocean mining, there is a crucial dilemma: on the one hand, sustaining the 'transition' to a more sustainable, low-carbon economy requires significant additional quantities of minerals and metals, which only the mining industry can provide, possibly also exploiting the extensive underwater deposits; on the other hand, the oceans are huge, almost unknown areas of the planet, with very delicate ecological balances and a biodiversity that is extremely vulnerable to small environmental changes. What would happen if, after overcoming the technological difficulties that still exist to make the exploitation cycle feasible, large remotely guided rovers, robots, and drones started to collect the enormous quantities of raw materials that exist on the seabed? The levels of uncertainty surrounding our knowledge of biotic and abiotic systems under the sea bed are perhaps still too great for us to be able to make definitive decisions.

However, until July 2020, ISA had approved thirty exploration contracts in international waters. Exploration does not automatically imply extraction; on the contrary, exploration activity often highlights the limitations and impracticality of mining activities. In any case, the frontier of deep sea/ocean mining in international waters (exploration in national waters is already underway—see Lusty & Murton, 2018) is approaching, still without being able to quantitatively assess the actual ecological

---

[30] https://www.isa.org.jm/. Accessed 29 March 2022.

[31] It is known by the acronym UNCLOS (United Nations Convention on the Law of the Sea): it is the international treaty that defines the rights and responsibilities of States in the use of the seas and oceans, setting out the guidelines that govern, among other things, the environment and the management of mineral resources.

risks (Jones et al., 2018) but only having a qualitative idea of the pros and cons of such activities, and especially without having internationally agreed upon the limits of environmental acceptability (Thompson, 2018).

The yield of deep-sea/ocean mining can be very high compared to land-based mining, and the problems of social acceptance of activities that take place in the darkness of the seabed are certainly negligible compared to those encountered in setting up mines on land. After all, who is interested in what might happen in a remote underwater area of the western Pacific Ocean, near the tiny Japanese Minami Torishima atoll? Here, on the seabed, about 5,000 m below the ocean surface, there are several metres of clayey muds containing an abundant marine fossil fauna from the Upper Eocene (34.4 million years ago) that is rich in rare earths. The area of the deposit is about 2,500 square kilometres and could provide 16 million tonnes of rare earth oxides, which would supply more than four times the world's current requirement of rare earths for hundreds of years (Ohta et al., 2020). At present, the technological difficulties and lack of significant economic returns to start mining at these great depths of the sea remain.[32]

The oceans are a great opportunity for the future, but at what price? We still need to understand it well, so that decisions are the result of shared responsible choices and not, as is often the case, the result of the risky enterprises of a few industrial lobbies.

References

Abrunhosa, M., Chambel, A., Peppoloni, S., & Chaminé, H. I. (Eds.). (2021). *Advances in Geoethics and Groundwater Management: Theory and practice for a sustainable development* (p. XLV+523). Springer. ISBN 978–3030593193. https://doi.org/10.1007/978-3-030-59320-9

Arvanitidis, N., Boon, J., Nurmi, P., & Di Capua, G. (2017). *White paper on responsible mining.* IAPG—International Association for Promoting Geoethics, http://www.geoethics.org/wp-responsible-mining. Accessed 29 March 2022.

Auffhammer, M. (2018). Quantifying economic damages from climate change. *Journal of Economic Perspectives, 32*(4), 33–52. https://doi.org/10.1257/jep.32.4.33

Bilham, N. (2020). Responsible mining and responsible sourcing of minerals: opportunities and challenges for cooperation across value chains. In G. Di Capua, P. T. Bobrowsky, S. W. Kieffer, & C. Palinkas (Eds.), *Geoethics: Status and future perspectives* (Special Publications 508, pp. 161–186). Geological Society of London. https://doi.org/10.1144/SP508-2020-130

Bonneuil, C., & Fressoz, J.-B. (2013). *L'Evénement Anthropocène - La Terre, l'histoire et nous* (p. 320). Seuil. ISBN 978-2021135008.

Boon, J. (2020a). Sociology for mineral exploration. In G. Di Capua, P. T. Bobrowsky, S. W. Kieffer, & C. Palinkas (Eds.), *Geoethics: Status and future perspectives* (Special Publications 508, pp. 149–159). Geological Society of London. https://doi.org/10.1144/SP508-2019-230

Boon, J. (2020b). *Relationships and the course of social events during mineral exploration—An applied sociology approach* (p. XIX+125). SpringerBriefs in Geoethics, Springer International Publishing. ISBN 978–3030379254. https://doi.org/10.1007/978-3-030-37926-1

---

[32] https://www.scientificamerican.com/article/mining-rare-earth-elements-from-fossilized-fish/. Accessed 29 March 2022.

Borgomeo, E. (2020). *Oro Blu – Storie di acqua e cambiamento climatico* (p. 176). Laterza. ISBN 978-8858140611.

Burns, W. C. G. (2011). Climate geoengineering: Solar radiation management and its implications for intergenerational equity. *Stanford Journal of Law, Science & Policy, 4*, 39–55.

Ceballos, G., Ehrlich, P. R., Barnosky, A. D., García, A., Pringle, R. M., & Palmer, T. M. (2015). Accelerated modern human–induced species losses: Entering the sixth mass extinction. *Science Advances, 1*(5), e1400253. https://doi.org/10.1126/sciadv.1400253

Ceballos, G., Ehrlich, P. R., & Dirzo, R. (2017). Biological annihilation via the ongoing sixth mass extinction signaled by vertebrate population losses and declines. *PNAS, 114*(30), E6089–E6096. https://doi.org/10.1073/pnas.1704949114

Chakravorty, A. (2019). Underground robots: How robotics is changing the mining industry. *EOS, 100.* https://doi.org/10.1029/2019EO121687

Chaminé, H. I., Abrunhosa, M., Barbieri, M., Naves, A., Errami, E., Aragão, A., & di Capua, G. (2021). Hydrogeoethics in sustainable water resources management facing water scarcity in Mediterranean and surrounding regions. *Mediterranean Geoscience Reviews, 3*, 289–292. https://doi.org/10.1007/s42990-021-00069-2

Cook, J., Supran, G., Lewandowsky, S., Oreskes, N., & Maibach, E. (2019). *America Misled: How the fossil fuel industry deliberately misled Americans about climate change.* George Mason University Center for Climate Change Communication. https://www.climatechangecommunication.org/america-misled/. Accessed 29 March 2022.

Crutzen, P. J. (2006). Albedo enhancement by stratospheric sulfur injections: A contribution to resolve a policy dilemma? *Climatic Change, 77*, 211–219. https://doi.org/10.1007/s10584-006-9101-y

Dal Corso, J., Bernardi, M., Sun, Y., Song, H., Seyfullah, L. J., Preto, N., Gianolla, P., Ruffell, A., Kustatscher, E., Roghi, G., Merico, A., Hohn, S., Schmidt, A. R., Marzoli, A., Newton, R. J., Wignall, P. B., & Benton, M. J. (2020). Extinction and dawn of the modern world in the Carnian (Late Triassic). *Science Advances, 6*(38), eaba0099. https://doi.org/10.1126/sciadv.aba0099

Dembe, A. E., & Boden, L. I. (2000). Moral hazard: A question of morality? *New Solutions, 10*(3), 257–279. https://doi.org/10.2190/1GU8-EQN8-02J6-2RXK

Gerten, D., Heck, V., Jägermeyr, J., Bodirsky, B. L., Fetzer, I., Jalava, M., Kummu, M., Lucht, W., Rockström, J., Schaphoff, S., & Schellnhuber, H. J. (2020). Feeding ten billion people is possible within four terrestrial planetary boundaries. *Nature Sustainability, 3*, 200–208. https://doi.org/10.1038/s41893-019-0465-1

Gibb, R., Redding, D. W., Chin, K. Q., Donnelly, C. A., Blackburn, T. M., Newbold, T. M. & Jones, K. E. (2020). Zoonotic host diversity increases in human-dominated ecosystems. *Nature, 584*, 398–402. https://doi.org/10.1038/s41586-020-2562-8

Groenfeldt, D. (2020). Ethical considerations in managing the hydrosphere: an overview of water ethics. In G. Di Capua, P. T. Bobrowsky, S. W. Kieffer, & C. Palinkas (Eds.), *Geoethics: Status and future perspectives* (Special Publications 508, pp. 201–2012). Geological Society of London. https://doi.org/10.1144/SP508-2020-99

IEA. (2019). *The Future of Hydrogen—Seizing today's opportunities* (p. 203). Report prepared by the IEA for the G20, Japan. International Energy Agency. https://www.iea.org/reports/the-future-of-hydrogen. Accessed 29 March 2022.

IPBES. (2018). *Summary for policymakers of the assessment report on land degradation and restoration of the Intergovernmental SciencePolicy Platform on Biodiversity and Ecosystem Services* (R. Scholes, L. Montanarella, A. Brainich, N. Barger, B. ten Brink, M. Cantele, B. Erasmus, J. Fisher, T. Gardner, T. G. Holland, F. Kohler, J. S. Kotiaho, G. Von Maltitz, G. Nangendo, R. Pandit, J. Parrotta, M. D. Potts, S. Prince, M. Sankaran and L. Willemen, Eds., p. 44). IPBES secretariat. https://ipbes.net/system/tdf/spm_3bi_ldr_digital.pdf?file=1&type=node&id=28335. Accessed 29 March 2022.

Izhitskiy, A. S., Zavialov, P. O., Sapozhnikov, P. V., Kirillin, G. B., Grossart, H. P., Kalinina, O. Y., Zalota, A. K., Goncharenko, I. V., & Kurbaniyazov, A. K. (2016). Present state of the Aral

Sea: Diverging physical and biological characteristics of the residual basins. *Scientific Reports, 6*, 23906. https://doi.org/10.1038/srep23906

Jones, D. O. B., Amon, D. J., & Chapman, A. S. A. (2018). Mining deep-ocean mineral deposits: What are the ecological risks? *Elements, 14*(5), 325–330. https://doi.org/10.2138/gselements.14. 5.325

Kartha, S., Kemp-Benedict, E., Ghosh, E., & Nazareth, A. (2020). *The Carbon Inequality Era—An assessment of the global distribution of consumption emissions among individuals from 1990 to 2015 and beyond*. Joint Research Report by Stockholm Environment Institute. https://oxfam.app. box.com/s/q36ywh37ppur8gl276zwe8goqr6utkej. Accessed 29 March 2022.

Keith, D. W. (2000). Geoengineering the climate: History and prospect. *Annual Review of Energy and the Environment, 25*, 245–284. https://doi.org/10.1146/annurev.energy.25.1.245

Keith, D. W. (2009). Engineering the planet. In S. Schneider & M. Mastrandrea (Eds.), *Climate Change Science and Policy* (Chapter 49, pp. 494–501). Island Press.

Kelman, I. (2019). Axioms and actions for preventing disasters. *Progress in Disaster Science, 2*, 100008. https://doi.org/10.1016/j.pdisas.2019.100008

Kiehl, J. T. (2006). Geoengineering climate change: Treating the symptom over the cause? *Climatic Change, 77*, 227–228. https://doi.org/10.1007/s10584-006-9132-4

Lamperti, F., Bosetti, V., Roventini, A., & Tavoni, M. (2019). The public costs of climate-induced financial instability. *Nature Climate Change, 9*, 829–833. https://doi.org/10.1038/s41558-019-0607-5

Lèbre, E., Owen, J. R., Corder, G. D., Kemp, D., Stringer, M., & Valenta, R. K. (2019). Source risks as constraints to future metal supply. *Environmental Science & Technology, 53*(18), 10571–10579. https://doi.org/10.1021/acs.est.9b02808

Lewis, S. L., & Maslin, M. A. (2018). *The human planet: How we created the Anthropocene* (p. 480). Pelican. ISBN 978-0241280881.

Lovelock, J. E. (1990). Hands up for the Gaia hypothesis. *Nature, 344*, 100–102. https://doi.org/10. 1038/344100a0

Lovelock, J. (2006). *La rivolta di Gaia* (p. 240). Rizzoli. ISBN 978-8817013145.

Lusty, P. A. J., & Murton, B. J. (2018). Deep-ocean mineral deposits: Metal resources and windows into earth processes. *Elements, 14*(5), 301–306. https://doi.org/10.2138/gselements.14.5.301

Lynas, M., Houlton, B. Z., & Perry, S. (2021). Greater than 99% consensus on human caused climate change in the peer-reviewed scientific literature. *Environmental Research Letters, 16*, 114005. https://doi.org/10.1088/1748-9326/ac2966

Mancheri, N. A., Sprecher, B., Bailey, G., Ge, J., & Tukker, A. (2019). Effect of Chinese policies on rare earth supply chain resilience. *Resources, Conservation and Recycling, 142*, 101–112. https:// doi.org/10.1016/j.resconrec.2018.11.017

Maximillian, J., Brusseau, M. L., Glenn, E. P., & Matthias, A. D. (2019). Chapter 25—Pollution and environmental perturbations in the global system. In M. L. Brusseau, I. L. Pepper, & C. P. Gerba (Eds.), *Environmental and pollution science* (3rd ed., pp. 457–476). https://doi.org/10. 1016/B978-0-12-814719-1.00025-2

McLean, M. R. (2020). Reaching out from Earth to the stars. In G. Di Capua, P. T. Bobrowsky, S. W. Kieffer, & C. Palinkas (Eds.), *Geoethics: Status and future perspectives* (Special Publications 508, pp. 297–302). Geological Society of London. https://doi.org/10.1144/SP508-2020-16

Meneganzin, A., Pievani, T., & Caserini, S. (2020). Anthropogenic climate change as a monumental niche construction process: Background and philosophical aspects. *Biology & Philosophy, 35*, 38. https://doi.org/10.1007/s10539-020-09754-2

Millennium Ecosystem Assessment. (2005). *Ecosystems and human well-being: Synthesis*. Island Press. http://www.bioquest.org/wp-content/blogs.dir/files/2009/06/ecosystems-and-hea lth.pdf. Accessed 29 March 2022.

Morini, S. (2014). *Il rischio: da Pascal a Fukushima* (p. 114). Bollati Boringhieri. ISBN 978-8833925035.

Mudd, G. (2020). Sustainable/responsible mining and ethical issues related to the Sustainable Development Goals. In G. Di Capua, P. T. Bobrowsky, S. W. Kieffer, & C. Palinkas (Eds.), *Geoethics:*

Status and future perspectives (Special Publications 508, pp. 187–199). Geological Society of London. https://doi.org/10.1144/SP508-2020-113

Munich Re. (2019). *Annual Report (Group) 2018* (p. 192). München. https://www.munichre.com/content/dam/munichre/contentlounge/website-pieces/documents/302-09122.pdf/_jcr_content/renditions/original.media_file.download_attachment.file/302-09122.pdf. Accessed 29 March 2022.

Oreskes, N. (2014). The scientific consensus on climate change. *Science, 306*(5702), 1686. https://doi.org/10.1126/science.1103618

Oreskes, N., & Conway, E. M. (2010). *Merchants of doubt: How a handful of scientists obscured the truth on issues from tobacco smoke to global warming* (p. 368). Bloomsbury Publishing. ISBN 978-1596916104.

Ohta, J., Yasukawa, K., Nozaki, T., Takaya, Y., Mimura, K., Fujinaga, K., Nakamura, K., Usui, Y., Kimura. J.-I., Chang, Q., & Kato, Y, (2020). Fish proliferation and rare-earth deposition by topographically induced upwelling at the late Eocene cooling event. *Scientific Reports, 10*, 9896. https://doi.org/10.1038/s41598-020-66835-8

Peppoloni, S. (2014). *Convivere con i rischi naturali* (p. 148). Il Mulino. ISBN 978-8815250780.

Preston, C. J. (2013). Ethics and geoengineering: Reviewing the moral issues raised by solar radiation management and carbon dioxide removal. *Wires Climate Change, 4*, 23–37. https://doi.org/10.1002/wcc.198

Ripple, W. J., Wolf, C., Newsome, T. M., Barnard, P., & Moomaw, W. R. (2020). World scientists' warning of a climate emergency. *BioScience, 70*(1), 8–12. https://doi.org/10.1093/biosci/biz088

Ripple, W. J., Wolf, C., Newsome, T. M., Gregg, J. W., Lenton, T. M., Palomo, I., Eikelboom, J. A. J., Law, B. A., Huq, S., Duffy, P. B., & Rockström, J. (2021). World scientists' warning of a climate emergency 2021. *BioScience, 71*(9), 894–898. https://doi.org/10.1093/biosci/biab079.

The White House. (1965). *Restoring the quality of our environment—Report of the Environmental Pollution Panel* (p. IX+293). President's Science Advisory Committee. U.S. Government Printing Office. http://ozonedepletiontheory.info/Papers/Revelle1965AtmosphericCarbonDioxide.pdf. Accessed 29 March 2022.

Thompson, J. F. H. (2018). Deep-ocean mineral resources. *Elements, 14*(5), 298. https://doi.org/10.2113/gselements.14.5.298

Truche, L., & Bazarkina, E. F. (2019). *Natural hydrogen the fuel of the 21st century* (E3S Web of Conferences 98, 03006, p. 5). https://doi.org/10.1051/e3sconf/20199803006

Ward, P. (2009). *The Medea hypothesis: Is life on earth ultimately self-destructive?* (p. 208) Princeton University Press. ISBN 978-0691130750.

United Nations. (2019). *World Population Prospects 2019: Highlights*. Department of Economic and Social Affairs, Population Division. ST/ESA/SER.A/423. https://population.un.org/wpp/Publications/Files/WPP2019_Highlights.pdf. Accessed 29 March 2022.

WEF. (2020). *The Global Risks Report 2020* (15th ed.). Insight Report. In partnership with Marsh & McLennan and Zurich Insurance Group. http://www3.weforum.org/docs/WEF_Global_Risk_Report_2020.pdf. Accessed 29 March 2022.

WWAP. (2019). *The United Nations World Water Development Report 2019: Leaving No One Behind*. UNESCO World Water Assessment Programme. UNESCO. https://en.unesco.org/themes/water-security/wwap/wwdr/2019. Accessed 29 March 2022.

Xu, C., Kohler, T. A., Lenton, T. M., Svenning, J.-M., & Scheffer, M. (2019). Future of the human climate niche. *PNAS, 117*(21), 11350–11355. https://doi.org/10.1073/pnas.1910114117

Zhou, B., Li, Z., & Chen, C. (2017). Global potential of rare earth resources and rare earth demand from clean technologies. *Minerals, 7*(11), 203. https://doi.org/10.3390/min7110203

Chapter 9
Geoethics for an Ecological Humanism

9.1 From the Emergence of an Ecological Conscience to Geoethics

In the 1940s, the American ecologist Aldo Leopold (see Chapter 2) coined the concept of *land ethics* (Leopold, 1949), which expresses the need to establish a new relationship between human communities and nature on a moral basis, identifying 'conservation' as the ethical criterion on which to base this relationship. Conserving ecological systems makes it possible to recover the state of harmony between human beings and the natural environment. This vision influences the cultural circles of that time in the United States and initiates, in the Western world, an ethical reflection on the environment and the development of a debate on the correct way for human beings to interact with the Earth system.

Numerous environmental movements emerged in the following decades, particularly since the 1970s, with a perspective mainly oriented towards the protection of the biosphere, although with important differences in approach, the result of different ecological visions. Alan Marshall from New Zealand groups the different visions that have developed over the last forty years into three categories: conservation ethics, ecological extension and libertarian extension (Marshall, 1993).

Conservation ethics, based on an anthropocentric view, argues that nature must be conserved and managed within a subordinate relationship to the needs of humanity. In other words, the management and use of the planet's natural resources are necessary to achieve satisfactory standards for human life, whose rights are put before any other living form. The *ecological extension* proposes a holistic view of the planet, recognising, as a fundamental element, the interdependence of all biotic and abiotic entities that make up the ecosystems and emphasising the value and preciousness of biodiversity. All the parts are dependent on mutual well-being, and recognition of the value of the whole comes before the rights of the individual parts. Finally, the *libertarian extension* proposes to extend equal rights also to the 'non-human' subjects that make up the Earth system and, therefore, also to the abiotic elements.

© The Author(s), under exclusive license
to Springer Nature Switzerland AG 2022
S. Peppoloni and G. Di Capua, *Geoethics*,
https://doi.org/10.1007/978-3-030-98044-3_9

This vision, which is part of a historical process of progressive affirmation of rights, would come, due to an excess of guarantees, to severely limit or even prohibit human actions on Earth in order to guarantee rights also to plants or microbes, as well as to everything abiotic.

From this picture, the complexity of the positions taken in the field of environmental ethics emerges, which can also be based on diametrically opposed concepts. Some positions consider human beings to be at the centre of the universe and the purpose of the agency, while others consider them to be the evil that oppresses Earth.

Leaving aside the different nuances of each conception, traditional anthropocentrism generally assigns humans a central and dominant position over nature, going so far as to consider, in an extreme vision, the value of nature only in relation to its usefulness for the human species. Thus, its protection and conservation would be independent of its intrinsic value and would depend on the fact that nature is functional to the well-being of humankind. In biocentrism, nature has a value in itself, independent of humans, who are considered living beings of equal importance to the others on Earth and who are obliged to protect and conserve nature to protect themselves, in the need to find a balance that guarantees their survival. Ecocentrism also goes beyond the biocentric position, attributing an intrinsic value to nature as the totality of what constitutes it, as a system of relationships. Human beings and all living beings in general are considered an integral part of nature and inseparable from it, and the value of nature as a whole is considered to be greater than the value of each organism considered individually. Ecocentrism moves the human being from a position of centrality with respect to nature, which determines our sense of domination over nature itself, to a position of a peer among other peers. Such a conception goes so far as (in the most extreme positions) to justify even self-elimination of the human species, if this serves to ensure the survival of the entire nature. The biocentric and ecocentric positions would guarantee attitudes and, therefore, actions by humanity that are respectful of nature, recognizing its full existential dignity (Rolston, 1998). Especially in the case of ecocentrism, this point of view seems completely incompatible with the current economic systems and global social organization (Peppoloni and Di Capua 2021a).

Finally, there is also the geocentric position, which can be understood as a further systemic extension of the ecocentric positions. Earth has an objective, self-produced systemic value which does not, therefore, depend on those species that can recognize or attribute that value to it. It is a value that Earth has by virtue of its relationships, value that is higher than that of a single species and single ecosystems that constitute it, as each part of the planetary system has reason to exist as part of the relationships constituting the whole (Rolston, 1998).

Each of these different visions contains important elements of truth for geoethical thought. The anthropocentric view places human beings at the centre of reality, as creator and agent of their sensory and rational experience, committed to ensuring their own survival and material and spiritual well-being. On the other hand, biocentrism captures the value of nature in itself, which is also fundamental to recognising the value of nature within us humans, as respect for nature inevitably passes through respect for ourselves as human beings. Finally, ecocentrism grasps the sense of the whole, of the connection between the parts, even if the inclusion of *anthropos* in the whole might be difficult to accept.

As stated in Peppoloni and Di Capua (2021a), placing the human at the centre of the geoethical vision of the world inevitably recalls the concept of anthropocentrism (Kopnina et al., 2018). It should be clarified that, as already stated in Peppoloni and Di Capua (2017, 2021b) and Peppoloni et al. (2019), anthropocentrism in geoethics is criticized in its traditional meaning and is reformulated in light of the principle of responsibility, as an ethical criterion for the agency. The concept of anthropocentrism intended as a position of prevarication of human interest on the right to existence of any other living and non-living entity is completely rejected by geoethics. It is contrary for geoethics to attribute to nature only an instrumental/functional value for our species, or to think that nature can be managed according to a relationship of subordination with respect to the needs of humanity. In the anthropocentric vision *stricto sensu*, nature has neither status nor value in herself. Consequently, the main function of the planet's natural resources is to ensure satisfactory standards of living for the human being, whose rights are placed before those of any other living entity. Usually, the development of the current economic, political, social, and cultural paradigms that have led to the over-exploitation of natural resources and the great inequalities between the rich and the poor people of the planet is traced back to anthropocentrism. Anthropocentrism would therefore underlie predatory capitalism and, in some ways, would support the most selfish part of the human being, almost justifying them, and leading them to even perform petty acts towards their fellows and towards what is other than self. This negative vision would leave no other possibility than to embrace different positions, such as those conceived by biocentrism, ecocentrism, or geocentrism.

Anthropocentrism, biocentrism, and ecocentrism (with its geocentric extension) are positions of environmental ethics (Hourdequin, 2015) which have been given very complex meanings. Furthermore, some adjustments and further specifications in the definitions have been made over time, for example, a strong anthropocentrism has been defined as opposed to a weak one (Norton, 1984). These specifications appear as attempts to overcome the rigidities of the different positions which do not really fit separately the human complexity and the complexity of the relationship between human beings and nature. In any case, our conviction remains that all these positions descend from an inevitable anthropocentrism of species; in other words, from a perception of things that, for us, as humans can only be anthropocentric, namely referring to the position that human beings give to themselves in relation to other than themselves. As human beings, we cannot fail to have an anthropocentric view or, as affirmed by Viola (1995), an anthropological point of view. Furthermore, biocentric and ecocentric positions are anthropocentric concepts since they are developed and expressed by the human being. They are symbolic representations of our perception of a world of relationships to which we assign an ethical meaning.

Conceiving anthropocentrism in these terms, i.e., referring to the inevitable perception that the human species has of its position on Earth is not in contradiction with being respectful of nature (Passmore, 1974) and acting responsibly towards it, having understood that we are an integral part of it and that, by protecting nature, we also safeguard ourselves.

Unfortunately, anthropocentrism, biocentrism, and ecocentrism have become terms in strong contraposition. In the case of biocentrism and ecocentrism, for example, the specificity of every living species, including the human species, is not considered as a given reality, but there is the tendency to amalgamate the peculiarities of each individual in a single apparently holistic dimension, without realizing that this operation is intrinsically anthropocentric.

Moreover, there is another important aspect to be considered, that these positions can also induce a deep sense of guilt in the human being, regardless of the responsibilities of the present and the opinionated attitude of looking at past history with the sensitivity and moral frameworks of the current time. The sense of guilt does not allow clarity of action but risks acting as a superstructure that, without being realised, it directs our choices, thus questioning the possibility of acting ethically, scientifically, and with common sense. The final result is that the sense of guilt feeds that feeling, strongly anthropocentric, which leads us to perceive ourselves as the rescuer of the environment, the planet, the cosmos. This highlights once again the persistent dichotomy between humans and nature, which is dominant in Western cultures and has spread to other human cultures, that ontological fracture that comes from afar (Capra, 1975, pp. 20–21; Morris, 2013) and that we try to hide by using words of elusive definitions.

We think a new attitude is needed. Environmental ethics has explored possible ways of relating to nature and has synthesized them in positions that have entered into conflict with each other (Kopnina 2018; Passmore, 1974), sparking endless discussions that have, in fact, created obstacles on the operational level that we have to remove quickly if we want to give common answers to global problems.

Geoethics seeks to go beyond the contrapositions and make a synthesis, incorporating the concepts of anthropocentrism, biocentrism, and ecocentrism/geocentrism in a unitary vision, which saves the best intuitions of the categories of environmental ethics and uses them to develop new pathways, more linked to reality, more effective on a practical level. To do this, it is necessary to recall two points of utmost importance in this synthesis process (Peppoloni and Di Capua, 2021a):

- Nature does not need humans to regenerate, change, and evolve. Human beings are not central to natural architecture, as the deep time of geology demonstrates, but they can legitimately build their living space, like other living species. Human species can recognize and ensure dignity also to what belongs to other species and to what is non-living. Additionally, this gives humans a lot of responsibility towards everything that is other than them.
- It is necessary to distinguish between the perception that humans can have of themselves, with respect to everything that is other than them, and their concrete actions being in relation to what is external to them. Geoethics admits that it is not possible for humankind to leave the anthropocentric point of view, intended as referred to the perceived position on Earth. The biocentric and ecocentric visions themselves are not exempt from an anthropocentric perspective since they were thought of

and argued by people. However, in implementation, it must be acknowledged that humans try to change reality on the basis of their needs and expectations. It is necessary to be aware that primary needs and human expectations (sometimes secondary or induced needs) do not coincide, otherwise the predatory aspect of our human nature (which in any case exists and feeds on expectations) can prevail over the part of us that seeks respect for and harmony with nature, that part that can instead find balance with the satisfaction of basic human needs.

The human being has always been a modifier of the natural system, to the extent that today it is no longer possible to conceive of a 'de-humanised' Earth system. It is no coincidence that we speak of social-ecological systems as constituents of the Earth system. Natural ecosystems have been modified, more or less consciously, by humans for at least 12,000 years (Ellis et al., 2021), certainly since the beginning of agriculture in the Neolithic period for Ruddiman (2003, 2007). If we look at the anthropogenic impact on the chemical composition of the atmosphere, certified by continuous measurements over several decades[1] (Ripple et al., 2020), or that of plastics now found within recent geological layers (Ross, 2018), we can understand that we have long since entered a phase of active modification of the planet, whereby any terrestrial environment is, in some way, 'altered' by anthropogenic action. Such manipulation, with more evident effects on the biosphere, would be traced back to the Pleistocene, when *Homo sapiens* is thought to have contributed to or even to have been solely responsible for the extinction of the so-called megafauna in many parts of the planet (Lewis & Maslin, 2018) and even backdated a few million years to the Pliocene, in this case with reference to hominids, as hypothesised by Faurby et al. (2020).

The idea of a prehistoric humanity in perfect harmony with nature would therefore seem to be unreliable and only found in isolated groups of Indigenous peoples who remained alien to the cultural and technological development of most human communities. Nevertheless, the concept of life of these isolated human groups and their relationship with the natural environment is striking, whereby nature is not perceived as something other than oneself, but is indivisible from the human, animal, and supernatural world. The Museum of Anthropology in Vancouver, Canada offers a touching narrative of the life view and relationship to the natural environment of the ancient native populations of British Columbia. They are still able to pass on to each other that connection with soil, water, and air that is the basis of their identity and so deep that the word 'nature' does not exist in the *Haida* language.

However, such forms of identification with nature may not necessarily represent the preservation of an archaic *status*, of an existential mode originally common to all humans and now lost, but rather evolutionary points of arrival. The geographer Reclus himself, when he states that 'man is nature becoming conscious of itself' (see Chapter 2), would capture the historical sense of this evolutionary process, which would progressively lead human beings towards the existential conception developed by those Indigenous communities. On the other hand, one cannot naively

---

[1] https://www.esrl.noaa.gov/gmd/ccgg/trends/mlo.html. Accessed 29 March 2022.

underestimate the fact that science and technology are by now inherent to the development of human civilisation which, therefore, cannot convert to forms of degrowth (Nørgård, 2013) or to come back living by re-adopting economic forms and structures used in past historic phases. Modern societies are too complex for one to think of being able to live like North American Indians or Australian Aborigines, outside of a historical-cultural perspective and a cumulative culture that cannot be forgotten or deconstructed. The great challenge of modern times is to grasp the meaning of that symbiotic living of Indigenous communities in order to achieve responsible and sustainable living (Conversi, 2021), in light of scientific knowledge and modern technological advances.

This challenge entails abandoning the idea that there exists an uncontaminated, wild and de-humanised nature ('The myth of a wilderness without humans'[2]), instead in favour of an idea of nature that is irreversibly anthropised (which does not automatically mean polluted or spoiled). It is on this idea of nature, made up of 'anthromes', or human biomes, originated by the direct interaction of human beings with ecosystems (Ellis & Ramankutty, 2008) which are self-regulating (Ellis & Haff, 2009), products of the act of awareness of a responsibilised humanity, that it will be possible to build a world completely different from the current one. This historical-technological-cultural process seems to now be unstoppably underway, even if it is still in a preliminary phase.

There remain concerns about a humanity that, while becoming increasingly globalised and technologised, is not yet firmly on the road to the cultural and moral development that must sustain and accompany technological progress. Without a shared ethics, a geoethics, it is not possible to structure trajectories of human development based on harmony with the being that permeates natural environments, on the appreciation and aesthetic and emotional enjoyment of the variety of nature, on the recognition of its necessity within the general framework of human existence, and on coherent and all-embracing respect for oneself and for every element that is other than oneself.

Geoethics grasps the profound meaning of anthropocentric, biocentric and ecocentric/geocentric positions and synthesises them in a vision that can be defined as 'ecological humanism' (Peppoloni & Di Capua, 2020), which corresponds to the concept of 'regenerated humanism' or 'planetary humanism' proposed by Morin (2020, pp. 105–114) and which rejects the quasi-divinisation of a human being dominating nature. Ecological humanism rejects the traditional anthropocentric conception of a human being dominating nature, to open the vision to what Dubos defined as an 'enlightened anthropocentrism' (Dubos, 1972).

The vision of the ecological humanism can only be centred on the human agent but in full awareness of the partiality and relativity of their rational, sensitive, and emotional experience, albeit within the richness of the manifestations of universal being. *Anthropos* is given the unconditional responsibility of being part of a whole and equal to all (Peppoloni & Di Capua, 2020), that same *anthropos* that Morin

---

[2] https://thereader.mitpress.mit.edu/the-myth-of-a-wilderness-without-humans/. Accessed 29 March 2022.

(2020, p. 106) considers *sapiens* and *demens*, *faber* and *mythologicus*, *oeconomicus* and *ludens*, in other words, *Homo complexus*.

9.2 The Question of the Anthropocene

The extent to which humanity today conditions Earth's global processes is reflected in the, as yet unformalised, proposal to introduce a new epoch into the scale of geological time dominated by anthropogenic action: the Anthropocene (Crutzen, 2002; Zalasiewicz et al., 2008; see the first paragraph of Chapter 2). This notion is contested and criticised by some scholars on both geological (Finney & Edwards, 2016) and philosophical grounds (Cuomo, 2017). From a scientific point of view, the debate is crisp and recently it was enriched by a new proposal for considering the Anthropocene an event in the geologic history of Earth and not a formal epoch. This would permit us to overcome some geological constraints in defining stratigraphic boundaries of Anthropocene, giving the possibility to define it 'in a way that is functional both to the international geological community and to the broader fields of environmental and social sciences. Formally defining the Anthropocene as a chronostratigraphical series and geochronological epoch with a precise global start date would drastically reduce the Anthropocene's utility across disciplines. Instead, we propose the Anthropocene be defined as a geological event, thereby facilitating a robust geological definition linked with a scholarly framework more useful to and congruent with the many disciplines engaging with human–environment interactions. Unlike formal epochal definitions, geological events can recognize the spatial and temporal heterogeneity and diverse social and environmental processes that interact to produce anthropogenic global environmental changes. Consequently, an Anthropocene Event would incorporate a far broader range of transformative human cultural practices and would be more readily applicable across academic fields than an Anthropocene Epoch, while still enabling a robust stratigraphic characterization' (Gibbard et al., 2021).

But, beyond its scientific certification, the cultural value of the concept of 'Anthropocene' is undeniable and the heated debates around it can be considered reflections on the human condition in times of deep and rapid anthropogenic global changes (Bohle & Bilham, 2019; Ehlers & Krafft, 2006; Latour, 2004, 2014, 2017), triggering unexplored and fruitful interactions between geosciences, social sciences, and philosophy (Bauer & Ellis, 2018; Braje & Lauer, 2020; Clark & Gunaratnam, 2017; Clark & Yusoff, 2017; Hamilton & Grinevald, 2015).

In their careful reflection about human impacts on the Earth system and on the social and cultural implications of today's debate on global warming, Bonneuil and Fressoz (2013) break down the Anthropocene into an articulated and varied complex of anthropogenic actions and interactions between human societies and nature, which are defined as the Thermocene (or even the Anglocene), the Tanatocene, the Phagocene, the Fronocene, the Agnotocene, the Capitalocene, and the

Polemocene, since 'in order to fruitfully embrace the contribution of the Anthropocene concept, we must also learn to distrust the grand narrative that accompanies it, to pass it through the sieve of criticism. That's how it works in science, history and democracy'.

Other authors have gone so far as to speak of the Sinforocene, defining it the epoch of calamities, 'to underline the disastrous climate excesses that have begun to appear more insistently since 2003' (Coccioni, 2008). Others talk about the Plasticene (Ross, 2018), evaluating the possible use of plastics dispersed in the environment, a sort of 'technofossils', as a geological marker of the Anthropocene (Zalasiewicz et al., 2016). Consider that global production of plastic resins and fibres rose from 2 million tonnes in 1950 to 380 million tonnes in 2015, and it is estimated that by 2050 there will be around 12,000 million tonnes of plastic waste in the environment (Geyer et al., 2017). A thin film of plastic could cover the entire surface of the earth (Ross, 2018). Plastics are present in water and ocean floors and are by now considered components of new anthropogenic rocks called 'plastiglomerates', a mixture of burnt plastics, beach sediments, volcanic rock fragments and organic debris (Corcoran et al., 2014).

However, if the debate on whether and how to identify a possible geological epoch of the Anthropocene is open (Head et al., 2021; Lewis & Maslin, 2018; Waters et al., 2018; Zalasiewicz et al., 2017), equally debated among its promoters is the lower chronological limit that would establish the end of the previous geological epoch (the Holocene) and the beginning of the Anthropocene. Some authors speak of a sort of 'Early Anthropocene' that started about 8000 years ago (Ruddiman, 2003, 2007), and others propose 1610, the year when the peak of the global CO_2 decrease was reached due to the natural reforestation of tropical American areas after the genocide of Indigenous agricultural populations by Spanish conquerors (Lewis & Maslin, 2018). Still others believe that fallout from atmospheric nuclear explosions is now sealed in sediments and will be for millions of years, so that 16 July 1945, the date of the first atomic bomb explosion in the New Mexico desert, would be the date that symbolically certifies a change not only in human history, but also in the history of Earth (Sanchez-Cabeza et al., 2021; Zalasiewicz et al., 2015).

The debate on the Anthropocene is accompanied by controversial assessments of whether or not a so-called 'good Anthropocene' is possible (Bennett et al., 2016). On one hand, there is the optimism of some 'ecofuturist' scholars who embrace the term, convinced that human intelligence and technology will be able to solve problems and control natural systems (Asafu-Adjaye et al., 2015). On the other hand, there are more cautious and critical scholars, for whom the world has been so irreversibly altered by human intervention that we must do all we can to understand and minimise our impacts and build a future that more effectively meets the interconnected needs of people, communities, and ecosystems (Biro, 2015; Dalby, 2016; Pereira et al., 2018).

Regardless of these discussions, humanity must consider its significant 'action on natural environments, capable of altering the living fabric by impoverishing and artificialising it' (Bonneuil & Fressoz, 2013).

One cannot fail to notice that the debate on the Anthropocene is also taking on very heated tones, and that the polemic that accompanies some discussions is sometimes excessively fueled by ideological motivations. Those in favour of its scientific formalisation seem to be looking forward to this moment as a turning point for humanity, which would then no longer have an alibi for its inaction against planetary pollution and the wicked economic and environmental policies of the last three hundred years. The resistance of those who oppose this formalisation derives both from substantial scientific considerations which criticize the meagreness of the thickness of the possible stratotype that would technically constitute the geological transition from a pre-anthropocene to an anthropocene epoch, and from a certain annoyance with the fact that, at a given moment in the geological history of the planet, human history could enter, and that, therefore, the deep time of geology would be somehow 'contaminated' at a precise instant by the historical time of humanity. It cannot be excluded that the current debate may also contain a desire for recognition on the part of some scientists in favour, as well as a snobbish attitude of unwillingness to change their position on the part of some scientists against.

However, it is clear that the reasons for the inertia of today's societies in solving the complex problems of modernity are not related to the formalisation of the Anthropocene, nor would things change if, in the face of a scientifically sound proposal, the International Stratigraphic Commission gave the green light to the Executive Committee of the IUGS (International Union of the Geological Sciences) to ratify an epoch or an event called 'Anthropocene'. The real challenge is to create a culture of knowledge, respect, and responsibility within human societies, for which the geosciences must also take responsibility. At this moment in history, the main danger could be a shift of responsibility from actions to definitions, from meaning to form for its own sake. Even turning our gaze away from Earth and looking at space as a possible location for resources may seem like a way to avoid the responsibility of acting here on Earth in terms of sustainability and environmental protection. Beyond the value and importance of space exploration for the advancement of scientific and technical knowledge, and the consequent contribution to improving the quality of our life on Earth, one is sometimes led to think that the great media interest in space enterprises is aimed at legitimising the huge outlays they need in public opinion and to build social consensus on possible mining in extraterrestrial places, a surreal debate, far removed from the real and compelling needs of humanity.

9.3 Ecological Humanism: Towards a Responsible Anthropocentric Vision

Any worldview that does not recognise the central role of anthropogenic impact on Earth's subsystems (atmosphere, hydrosphere, cryosphere, geosphere, biosphere, anthroposphere) (Crutzen, 2006; Ellis, 2017; Head et al., 2021; Ruddiman, 2013; Steffen et al., 2007; Steffen, Broadgate, et al., 2015; Steffen, Grinevald, et al., 2011; Steffen, Persson, et al., 2011; Steffen, Richardson, et al., 2015; Zalasiewicz et al., 2008) and, as a consequence, does not assign humanity the responsibility to take action to safeguard itself and the planet, highlights a profound cultural and material crisis of humans, particularly Western humans (Jonas, 2001). As noted by Morris (2013, p. 15),

> Hans Jonas describes the crisis he sees arising from 'the threat we pose to the planet's ecology,' one that forces us to look anew at 'one of the oldest philosophical questions, that of the relationship between human being and nature, between mind and matter - in other words, the age-old question of dualism.' Jonas sees the ecological crisis originating in unrestrained scientific and technological development occurring without an objective ethical framework to serve as a guide. Ethics lags behind action and consists of weak attempts to circumscribe the potentially negative consequences of actions already set in motion.

Hence, the inevitability that anthropogenic change leads back to an anthropocentric vision (Peppoloni & Di Capua, 2017), albeit one enlightened by the principle of responsibility (Jonas, 1979), to deal with this crisis.

Geoethics encourages society to become fully aware of such dynamics and the effects they are capable of triggering, providing a conceptual framework of reference to pursue safety and health for society, and taking on ecologically oriented concerns and sensitivities through ethical behaviours and practices (Peppoloni et al., 2019).

Tackling the global problems of current times requires action not only by scientists but by any human being (Passmore, 1974) who, for this purpose, must be made aware of the issues through correct information and adequate scientific education and be made responsible in relation to the actual contribution one can make to the community, to implement an ethics of shared responsibility (Lazzarini, 2006), even if, as Nollkaemper (2018) states, there is always the risk that 'the sharing of responsibility can lead to a diffusion of responsibility that makes it more difficult to determine who is responsible for what. Thereby, it can undermine the effectiveness of global governance and moreover generate a new set of responsibility gaps.'

However, it is only by redefining the centrality of the human being in the Earth system that it is possible to make them responsible for their actions and to give life to an ecological humanism, which is not limited to a perspective of survival of the human species, but which, through a scientifically founded 'geosophy', opens up the possibilities of *anthropos* to a future of authentic and conscious unity of nature.

9.4   A Systemic Vision and a Law Aimed at Educating

Geosciences are showing that Earth cannot be considered a simple agglomeration of parts but a complex adaptive system, of which we have knowledge affected by different levels of uncertainty and whose dynamics can be analyzed and predicted on a probabilistic basis.

This interpretation of natural reality can only progressively influence the future evolution of jurisprudence. The law will have to deal with the systemic dimension of issues, impacts, and actions undertaken for management of the environment, the defense against natural and anthropogenic risks, and the use of natural resources. But, at the same time, it will have to consider the counter-intuitive dimension of the observations made on too short time periods, which are insufficient to evaluate the extent of phenomena whose effects can occur on longer time intervals. Indeed, in some cases, those observations could show characteristics that, in the short term, contradict the long-term trend of local and global changes (Peppoloni, 2020).

The solutions on which to base the law in geoscience matters will have to show modularity in order to allow for an adaptation to the contingent and a potential implementation of norms that can extend over wider temporal and spatial scales. Thus, for example, if water management will have to fulfill the needs of the moment, sometimes in situations of emergency, it will not be able to avoid considering the projection of the effects in the years to follow and the possible impact on larger areas by virtue of the spatial continuity of the hydrogeological systems. Furthermore, the substantial non-linearity of natural processes will have to be considered, so a perturbative action of the ecological system will be able to follow incremental effects over time, capable of producing an irrecoverable loss of natural resources and services. This is the case of the depletion of coastal aquifers due to excessive pumping, which causes the salinization of the aquifers over time and compaction of the sediments that contain them, leading to irreversible reduction of their storage capacity of underground fresh water (Peppoloni, 2020).

Furthermore, it will be necessary to find ways of combining the law with the uncertainty of science and with analyses and results that inevitably have a probabilistic character. The precautionary principle (Bourguignon, 2015), often invoked to make decisions in case of incomplete knowledge of phenomena, is not consistent with the epistemological foundations (verifiability and falsifiability) of scientific methods and, although its formulation has varied over time, it actually justifies a vision which imposes constraints to action, to the point of preferring inaction rather than a conscious, responsible, and scientifically thoughtful action, although affected by degrees of uncertainty. A scientific approach to environmental issues must be able to evaluate the pros and cons of an action, short- and long-term impacts, considering that irreversibility is an intrinsic condition of nature. Consequently, the acceptability of the proposed solutions must respond to principles of prudence and foresight based on scientific knowledge (Peppoloni, 2020).

In any case, what has been said does not solve the question of societal acceptability of decisions, which instead should be provided for by legal systems as a codified way of confrontation with citizenship. Citizens must be informed about decisions that will affect their lives, their activities and the territories they inhabit.

Furthermore, they must be made aware of the benefits of a choice, as well as of the possible negative consequences that can occur. Ability to build a growth path with local populations is needed, which is shared also through listening to legitimate distrust and contrarieties, in order to dissipate the perception of inattention and irrelevance that spreads when works of great environmental impact must be realized, such as landfills, quarries, mines, oil extraction plants or large infrastructures. Making responsible citizens passes through scientific information, the capacity for social involvement, the simplification of communication without banalization, the definition of compensation mechanisms that are not only economic but also capable of improving the quality of environmental and educational assets and cultural activities of the territory (Peppoloni, 2020).

The law must continue to pragmatically maintain its regulatory, categorical and sanctioning root, which makes interventions against environmental crimes even more intransigent, to be configured as serious crimes against safety and human health, with projection also on future generations. However, at the same time, the law should give a more extensive meaning to the norm, also assuming an educational function for which the citizen is placed in a position to grasp the value before the rule, or better the value that is behind the rule. This could be achieved with the development of devices that define a framework of mandatory duties of citizens towards the common home (their territory). These would favor and reward good practices and virtuous behaviors, ensuring decisional action by governmental authorities based on rigorous scientific knowledge, which considers information and scientific education a duty that universities and research institutes must certify. This would counteract those who, by abusing the right to free expression, spread false or manipulated 'scientific' information, or pseudo-scientific theories capable of inducing behaviors that could compromise the safety of the population or the natural environment (Peppoloni, 2020).

As Capra and Mattei (2017) clearly state, 'Law is not a set of rules and principles written in books, accessible only to jurists, but it is alive and present, an expression of our social and ethical behavior, made of obligations one to another and toward the community. Only if law is felt in this way, it can be born and be generative.'

9.5 Charter for a Human Responsible Development

Geoethics is part of a process of historical development of thought and draws from it the numerous elements necessary to structure its vision, theoretical articulation, and operational praxis. If, for some contents and aspects, geoethics could be classified as one of the environmental ethics (Hourdequin, 2015), nevertheless its thought goes beyond their contrasting positions, proposing itself as a conceptual and value structure rooted in the peculiarities of the philosophy of geology (Frodeman, 1995), on which to build a new *modus vivendi et operandi* in the twenty-first century and beyond.

Geoethics is based on the human agent who, through the various dimensions of responsibility, takes on the duty to act in accordance with one's complex biological–emotional–rational nature, shaped by the knowledge of oneself and the world, and motivated by self-recognition to be a moral being. This awareness gives sense to human existence, placing human dignity and the responsibility for human action in every circumstance in the foreground.

Hence, the duty to guarantee to themselves and any element other than self the same value and opportunities, on the basis of conscious diversity. Ecological humanism opens up the possibility that human and nature rights are accompanied by binding duties of each human being, not as a jurisprudential dictation but as evidence of one's humanity (Peppoloni & Di Capua, 2020), not unlike Morin's vision (2020, p. 107), which points out personal reform as a necessary step to achieve an ethical revitalisation of the individual. This could be sanctioned through the proposal of an international Charter for Responsible Human Development (Peppoloni & Di Capua, 2020), in which it is recognized that each human being has the duty to:

(1) respect the freedom and social, cultural and economic rights of others, rejecting any type of discrimination and recognizing to each one dignity and freedom to develop their own personality, creative potential and talent;

(2) develop and exercise responsibility towards themselves, other individuals, the social structures to which he/she belongs, nature in its both animated and inanimate forms, contributing to the realization of inclusive, peaceful, resilient and sustainable human communities;

(3) improve one's own knowledge and preparation, within the limits of their possibilities, trying to draw the information useful for their training from sources (institutions, organizations, scientific and technical bodies) that guarantee scientific quality and accuracy;

(4) make their knowledge, competence and abilities available to others to contribute to a responsible and ecologically sustainable human development, which also guarantees future generations conditions of well-being, safety and self-determination;

(5) ensure inclusivity, equity, solidarity, justice, sustainability in their decisions and actions, taking care that they are based on knowledge that also includes the assessment of possible consequences;

(6) cooperate to build and defend socio-economic and political-legal systems that guarantee respect for human rights and the reduction of inequalities, which promote the intellectual, spiritual and material freedom and safety of the individual, without distinction of ethnicity, gender, language, religion, political opinion or economic and social condition;

(7) act with prudence and foresight towards the natural environment, considering that one's actions can have repercussions in space and time far beyond what is foreseeable, due to the intrinsic epistemic uncertainty of social-ecological systems;

(8) contribute, according to one's own possibilities, to the identification of a safe and healthy operating space for the human species, by using the resources of nature carefully, respecting their fair distribution and the planetary ecological boundaries;

(9) protect the environment from degradation, from any form of pollution and excessive exploitation, ensuring aesthetic quality of the living spaces, and harmony and balance between their constituent parts.

The current ecological crisis may be a great opportunity to improve ourselves, to live without fear. On this pathway, geoethical thinking can provide those ethical and scientific references that allow each person to adhere to one's own authenticity as a human being.

References

Asafu-Adjaye, J., Foreman, C., Pritzker, R., Blomqvist, L., Keith, D., Roy, J., Lewis, M., Brand, S., Sagoff, M., Brook, B., Lynas, M., Shellenberger, M., Defries, R., Nordhaus, T., Stone, R., Ellis, E., Pielke, R., & Teague, P. (2015). *An Ecomodernist Manifesto* (p. 31). Breakthrough Institute. http://www.ecomodernism.org/manifesto. Accessed 29 March 2022.
Bauer, A. M., & Ellis, E. C. (2018). The Anthropocene divide: Obscuring understanding of social-environmental change. *Current Anthropology, 59*(2), 209–227. https://doi.org/10.1086/697198
Bennett, E. M., Solan, M., Biggs, R., McPhearson, T., Norstrom, A. V., Olsson, P., Pereira, L., Peterson, G. D., Raudsepp-Hearne, C., Biermann, F., Carpenter, S. R., Ellis, E. C., Hichert, T., Galaz, V., Lahsen, M., Milkoreit, M., Lopez, B. M., Nicholas, K. A., Preiser, R., … Xu, J. (2016). Bright spots: Seeds of a good Anthropocene. *Frontiers in Ecology and the Environment, 14*(8), 441–448. https://doi.org/10.1002/fee.1309
Biro, A. (2015). The good life in the greenhouse? Autonomy, democracy, and citizenship in the Anthropocene. *Telos, 2015*(172), 15–37. https://doi.org/10.3817/0915172015
Bohle, M., & Bilham, N. (2019). The 'Anthropocene Proposal': A possible quandary and a work-around. *Quaternary, 2*(2), 19. https://doi.org/10.3390/quat2020019
Bonneuil, C., & Fressoz, J.-B. (2013). *L'Evénement Anthropocène - La Terre, l'histoire et nous* (p. 320). Seuil. ISBN 978-2021135008.
Bourguignon, D. (2015). *The precautionary principle—Definitions, applications and governance* (p. 28). EPRS—European Parliamentary Research Service, European Parliament, European Union. ISBN 978-9282384800. https://doi.org/10.2861/821468
Braje, T. J., & Lauer, M. (2020). A meaningful Anthropocene?: Golden spikes, transitions, boundary objects, and anthropogenic seascapes. *Sustainability, 12*(16), 6459, 1–12. https://doi.org/10.3390/su12166459
Capra, F. (1975). *The Tao of physics: An exploration of the parallels between modern physics and eastern mysticism* (p. 330). Shambhala. ISBN 0-87773-077-6.
Capra, F., & Mattei, U. (2017). *Ecologia del Diritto: Scienza, politica, beni comuni* (p. 256). Aboca Edizioni. ISBN 978-8898881413
Clark, N., & Gunaratnam, Y. (2017). Earthing the Anthropos? From 'Socialising the Anthropocene' to Geologising the Social. In G. Delanty (Ed.), *Agency and historical time: Social theory in the age of the anthropocene. 20th Anniversary Special Issue of the European Journal of Social Theory, 20*(1), 146–163.
Clark, N., & Yusoff, K. (2017). Geosocial Formations and the Anthropocene. *Theory, Culture & Society, Special Issue 'Geosocial Formations and the Anthropocene', 34*(2–3), 3–23.
Coccioni, R. (2008). I geologi del Sinforocene. *Geologia dell'Ambiente, 3*, 3–4.
Conversi, D. (2021). Exemplary ethical communities. A new concept for a liveable Anthropocene. *Sustainability, 13*(10), 5582. https://doi.org/10.3390/su13105582
Corcoran, P. L., Moore, C. J., & Jazvac, K. (2014). An anthropogenic marker horizon in the future rock record. *GSA Today, 24*(6), 4–8. https://doi.org/10.1130/GSAT-G198A.1
Crutzen, P. J. (2002). Geology of mankind. *Nature, 415*, 23. https://doi.org/10.1038/415023a
Crutzen, P. J. (2006). The "Anthropocene". In E. Ehlers & T. Krafft (Eds.), *Earth system science in the Anthropocene: Emerging issues and problems* (pp. 13–18). Springer. https://doi.org/10.1007/3-540-26590-2_3
Cuomo, C. J. (2017). The Anthropocene: Foregone or Premature Conclusion? Examining the ethical implications of naming a new epoch. *Earth: The Science Behind the Headlines, 10–11.* https://www.earthmagazine.org/article/comment-anthropocene-foregone-or-premature-conclusion-examining-ethical-implications-naming. Accessed 29 March 2022.
Dalby, S. (2016). Framing the Anthropocene: The good, the bad and the ugly. *Anthropocene Review, 3*(1), 33–51. https://doi.org/10.1177/2053019615618681
Dubos, R. (1972). *A God Within* (p. IX-325). Charles Scribner's Sons. ISBN 978-0684127682.
Ehlers, E., & Krafft, T. (2006). *Earth system science in the Anthropocene* (p. 273). Springer, Berlin. ISBN 978-3540265887.

Ellis, E. C. (2017). Physical geography in the Anthropocene. *Progress in Physical Geography, 41*(5), 525–532. https://doi.org/10.1177/0309133317736424

Ellis, E. C., Gauthier, N., Goldewijk, K. K., Bird, R. B., Boivin, N., Díaz, S., Fuller, D. Q., Gill, J. L., Kaplan, J. O., Kingston, N., Locke, H., McMichael, C. N. H., Ranco, D., Rick, T. C., Rebecca S. M., Stephens, L., Svenning, J. C., & Watson, J. E. M. (2021). People have shaped most of terrestrial nature for at least 12,000 years. *PNAS, 118*(17), e2023483118. https://doi.org/10.1073/pnas.2023483118

Ellis, E. C., & Haff, P. K. (2009). Earth science in the Anthropocene: New epoch, new paradigm. *New Responsibilities. EOS Transactions, 90*(49), 473. https://doi.org/10.1029/2009EO490006

Ellis, E. C., & Ramankutty, N. (2008). Putting people in the map: Anthropogenic biomes of the world. *Frontiers in Ecology and the Environment, 6*(8), 439–447. https://doi.org/10.1890/070062

Faurby, S., Silvestro, D., Werdelin, L., & Antonelli, A. (2020). Brain expansion in early hominins predicts carnivore extinctions in East Africa. *Ecology Letters, 23*, 537–544. https://doi.org/10.1111/ele.13451

Finney, S. C., & Edwards, L. E. (2016). The "Anthropocene" epoch: Scientific decision or political statement? *GSA Today, 26*(3), 4–10. https://doi.org/10.1130/GSATG270A.1

Frodeman, R. (1995). Geological reasoning: Geology as an interpretive and historical science. *Geological Society of America Bulletin, 107*(8), 960–968. https://doi.org/10.1130/0016-7606(1995)107%3c0960:GRGAAI%3e2.3.CO;2

Geyer, R., Jambeck, J. R., & Lavender Law, K. (2017). Production, use, and fate of all plastics ever made. *Science Advances, 3*(7), e1700782. https://doi.org/10.1126/sciadv.1700782

Gibbard, P. L., Bauer, A. M., Edgeworth, M., Ruddiman, W. F., Gill, J. L., Merritts, D. J., Finney, S. C., Edwards, L. E., Walker, M. J.C. Maslin, M., & Ellis, E. C. (2021). *A practical solution: The Anthropocene is a geological event, not a formal epoch.* Episodes, Online First. https://doi.org/10.18814/epiiugs/2021/021029

Hamilton, C., & Grinevald, J. (2015). Was the Anthropocene anticipated? *The Anthropocene Review, 2*(1), 59–72. https://doi.org/10.1177/2053019614567155

Head, M. J., Steffen, W., Fagerlind, D., Waters, C., Poirier, C., Syvitski, J., Zalasiewicz, J., Barnosky, A., Cearreta, A., Jeandel, C., Leinfelder, R., Mcneill, J. R., Rose, N., Summerhayes, C., Wagreich, M., & Zinke, J. (2021). *The Great Acceleration is real and provides a quantitative basis for the proposed Anthropocene Series/Epoch.* Episodes, Online First. https://doi.org/10.18814/epiiugs/2021/021031

Hourdequin, M. (2015). *Environmental ethics—From theory to practice* (p. 256). Bloomsbury Academic. ISBN 978-1472510983.

Jonas, H. (1979). *Das Prinzip Verantwortung: Versuch einer Ethik für die technologische Zivilisation.* Suhrkamp [*The imperative of responsibility: In search of ethics for the technological age* (translation of Das Prinzip Verantwortung) trans. Hans Jonas and David Herr (1979). ISBN 0-226-40597-4 (University of Chicago Press, 1984), ISBN 0-226-40596-6].

Jonas, H. (2001). *Dalla fede antica all'uomo tecnologico* (p. 496). Il Mulino. ISBN 978-8815083449.

Kopnina, H., Washington, H., Taylor, B., & Piccolo, J. (2018). Anthropocentrism: More than just a misunderstood problem. *Journal of Agricultural and Environmental Ethics, 31*, 109–127. https://doi.org/10.1007/s10806-018-9711-1

Latour, B. (2004). *Politics of nature—How to bring the sciences into democracy* (p. 320). Harvard University Press. ISBN 978-0674013476.

Latour, B. (2014). Agency at the time of the Anthropocene. *New Literary History, 45*(1), 1–18.

Latour, B. (2017). *Facing Gaia: Eight Lectures on the New Climatic Regime* (p. 327). Polity Pres. ISBN 978-0745684338.

Lazzarini, G. (2006). *Etica e scenari di responsabilità sociale* (p. 208). FrancoAngeli. ISBN 978-8846480989.

Leopold, A. (1949). *A Sand County Almanac: And Sketches Here and There* (p. 240). Oxford University Press, Enlarged edition (31 December 1968). ISBN 978-0195007770.

Lewis, S. L., & Maslin, M. A. (2018). *The human planet: How we created the Anthropocene* (p. 480). Pelican. ISBN 978-0241280881.

Marshall, A. (1993). Ethics and the extraterrestrial environment. *Journal of Applied Philosophy, 10*(2), 227–236.

Morin, E. (2020). *Changeons de voie: Les leçons du coronavirus* (p. 160). Éditions Denoël. ISBN 978-2207161869.

Morris, T. (2013). *Hans Jonas's ethic of responsibility: From ontology to ecology* (p. 246). State University of New York Press. ISBN 978-1438448817.

Nollkaemper, A. (2018). The duality of shared responsibility. *Contemporary Politics, 24*(5), 524–544. https://doi.org/10.1080/13569775.2018.1452107

Nørgård, J. S. (2013). Happy degrowth through more amateur economy. *Journal of Cleaner Production, 38*, 61–70. https://doi.org/10.1016/j.jclepro.2011.12.006

Norton, B. G. (1984). Environmental ethics and weak anthropocentrism. *Environmental Ethics, 6*(2), 131–148. https://doi.org/10.5840/ENVIROETHICS19846233

Passmore, J. (1974). *Man's responsibility for nature: Ecological problems and western traditions* (p. 213). Charles Scribner's Sons. ISBN 0-684138158.

Peppoloni, S. (2020). Geoscienze, geoetica e diritto. In M. Zanichelli (Ed.), *Il diritto visto da fuori: scienziati, intellettuali, artisti si interrogano sul senso della giuridicità oggi* (pp. 71–82). FrancoAngeli. ISBN 978-8835106845.

Peppoloni, S., & Di Capua, G. (2017). Geoethics: Ethical, social and cultural implications in geosciences. *Annals of Geophysics, 60, Fast Track 7: Geoethics at the heart of all geoscience.* https://doi.org/10.4401/ag-7473

Peppoloni, S., & Di Capua, G. (2020). Geoethics as global ethics to face grand challenges for humanity. In G. Di Capua, P. T. Bobrowsky, S. W. Kieffer, & C. Palinkas (Eds.), *Geoethics: Status and future perspectives* (Special Publications 508, pp. 13–29). Geological Society of London. https://doi.org/10.1144/SP508-2020-146

Peppoloni, S., & Di Capua, G. (2021a). Geoethics to start up a pedagogical and political path towards future sustainable societies. *Sustainability, 13*(18), 10024. https://doi.org/10.3390/su131810024

Peppoloni, S., & Di Capua, G. (2021b). Current definition and vision of geoethics. In M. Bohle & E. Marone (Eds.), *Geo-societal Narratives—Contextualising geosciences* (pp. 17–28). Palgrave Macmillan. https://doi.org/10.1007/978-3-030-79028-8_2

Peppoloni, S., Bilham, N., & Di Capua, G. (2019). Contemporary geoethics within the geosciences. In M. Bohle (Ed.), *Exploring geoethics—Ethical implications, societal contexts, and professional obligations of the geosciences* (pp. 25–70). Palgrave Pivot. https://doi.org/10.1007/978-3-030-12010-8_2

Pereira, L. M., Hichert, T., Hamann, M., Preiser, R., amp; Biggs, R. (2018). Using futures methods to create transformative spaces: Visions of a good Anthropocene in Southern Africa. *Ecology and Society, 23*(1), 19. https://doi.org/10.5751/ES-09907-230119

Ripple, W. J., Wolf, C., Newsome, T. M., Barnard, P., & Moomaw, W. R. (2020). World scientists' warning of a climate emergency. *BioScience, 70*(1), 8–12. https://doi.org/10.1093/biosci/biz088

Rolston, H. (1998). Challenges in environmental ethics. In M. E. Zimmerman, J. B. Callicott, G. Sessions, K. J. Warren, & J. Clark (Eds.), *Environmental Philosophy: From Animal Rights to Radical Ecology* (2nd ed., pp. 124–144). Prentice Hall.

Ross, N. L. (2018). The "Plasticene" Epoch? *Elements, 14*(5), 291. https://doi.org/10.2138/gselements.14.5.291

Ruddiman, W. F. (2003). The anthropogenic greenhouse era began thousands of years ago. *Climatic Change, 61*, 261–293.

Ruddiman, W. F. (2007). The early anthropogenic hypothesis: Challenges and responses. *Reviews of Geophysics, 45*, RG4001. https://doi.org/10.1029/2006RG000207

Ruddiman, W. F. (2013). The Anthropocene. *Annual Review of Earth and Planetary Sciences, 41*(1), 45–68. https://doi.org/10.1146/annurev-earth050212-123944

Sanchez-Cabeza, J.-A., Rico-Esenaro, S. D., Corcho-Alvarado, J. A., Röllin, S., Carricart-Ganivet, J. P., Montagna, P., Ruiz-Fernández, A. C., & Cearreta, A. (2021). Plutonium in coral archives: A good primary marker for an Anthropocene type section. *Science of the Total Environment, 771,* 145077. https://doi.org/10.1016/j.scitotenv.2021.145077

Steffen, W., Broadgate, W., Deutsch, L., Gaffney, O., & Ludwig, C. (2015). The trajectory of the Anthropocene: The great acceleration. *The Anthropocene Review, 2*(1), 81–98. https://doi.org/10.1177/2053019614564785

Steffen, W., Crutzen, P. J., & McNeill, J. R. (2007). The Anthropocene: Are humans now overwhelming the great forces of nature. *AMBIO: A Journal of the Human Environment, 36*(8), 614–621.

Steffen, W., Grinevald, J., Crutzen, P., & McNeill, J. (2011). The Anthropocene: Conceptual and historical perspectives. *Philosophical Transactions of the Royal Bibliography Society A: Mathematical, Physical and Engineering Sciences, 369*(1938), 842–867. https://doi.org/10.1098/rsta.2010.0327

Steffen, W., Persson, Å., Deutsch, L., Zalasiewicz, J., Williams, M., Richardson, K., Crumley, C., Crutzen, P., Folke, C., Gordon, L., Molina, M., Ramanathan, V., Rockström, J., Scheffer, M., Schellnhuber, H. J., & Svedin, U. (2011). The Anthropocene: From global change to planetary stewardship. *Ambio, 40,* 739–761. https://doi.org/10.1007/s13280-011-0185-x

Steffen, W., Richardson, K., Rockström, J., Cornell, S. E., Fetzer, I., Bennett, E. M., Biggs, R., Carpenter, S. R., de Vries, W., de Wit, C. A., Folke, C., Gerten, D., Heinke, J., Mace, G. M., Persson, L. M., Ramanathan, V., Reyers, B., & Sörlin, S. (2015). Planetary boundaries: Guiding human development on a changing planet. *Science, 347*(6223), 1259855–1259855. https://doi.org/10.1126/science.1259855

Viola, F. (1995). *Stato e Natura.* Edizioni Anabasi SPA. ISBN 88-41780169.

Waters, C. N., Zalasiewicz, J., Summerhayes, C., Fairchild, I. J., Rose, N. L., Loader, N. J., Shotyk, W., Cearreta, A., Head, M. J., Syvitski, J. P. M., Williams, M., Wagreich, M., Barnosky, A. D., An, Z., Leinfelder, R., Jeandel, C., Gałuszka, A., Ivar do Sul, J. A., Gradstein, F., Steffen, W., … Edgeworth, M. (2018). Global Boundary Stratotype Section and Point (GSSP) for the Anthropocene Series: Where and how to look for potential candidates. *Earth-Science Reviews, 178,* 379–429. https://doi.org/10.1016/j.earscirev.2017.12.016

Zalasiewicz, J., Waters, C. N., Ivar do Sul, J. A., Corcoran, P. L., Barnosky, A. D., Cearreta, A., Edgeworth, M., Gałuszka, A., Jeandel, C., Leinfelder, R., McNeill, J. R., Steffen, W., Summerhayes, C., Wagreich, M., Williams, M., Wolfe, A. P., & Yonan, Y. (2016). The geological cycle of plastics and their use as a stratigraphic indicator of the Anthropocene. *Anthropocene, 13,* 4–17. https://doi.org/10.1016/j.ancene.2016.01.002

Zalasiewicz, J., Waters, C. N., Williams, M., Barnosky, A. D., Cearreta, A., Crutzen, P., Ellis, E., Ellis, M. A., Fairchild, I. J., Grinevald, J., Leinfelder, R., McNeill, J., Poirier, C., Richter, D., Steffen, W., Vidas, D., Wagreich, M., Wolfe, A. P., & Zhisheng, A. (2015). When did the Anthropocene begin? A mid-twentieth century boundary level is stratigraphically optimal. *Quaternary International, 383,* 196–203. https://doi.org/10.1016/j.quaint.2014.11.045

Zalasiewicz, J., Waters, C. N., Summerhayes, C. P., Wolfe, A. P., Barnosky, A. D., Cearreta, A., Crutzen, P., Ellis, E., Fairchild, I. J., Gałuszka, A., Haff, P., Hajdas, I., Head, M. J., Ivar do Sul, J. A., Jeandel, C., Leinfelder, R., McNeill, J. R., Neal, C., Odada, E., … William, M. (2017). The working group on the Anthropocene: Summary of evidence and interim recommendations. *Anthropocene, 19,* 55–60. https://doi.org/10.1016/j.ancene.2017.09.001

Zalasiewicz, J., Williams, M., Smith, A. G., Barry, T. L., Coe, A. L., Bown, P. R., Brenchley, P., Cantrill, D., Gale, A., Gibbard, P., Gregory, F. J., Hounslow, M. W., Kerr, A. C., Pearson, P., Knox, R., Powell, J., Waters, C., Marshall, J., Oates, M., … Stone, P. (2008). Are we now living in the Anthropocene? *GSA Today, 18*(2), 4–8. https://doi.org/10.1130/GSAT01802A.1

Printed in the United States
by Baker & Taylor Publisher Services